Multi-Robot Systems - New Advances

Edited by Serdar Küçük

Published in London, United Kingdom

Multi-Robot Systems - New Advances
http://dx.doi.org/10.5772/intechopen.108092
Edited by Serdar Küçük

Contributors
Jirui Fu, Joon-Hyuk Park, Zubaidah Al-Mashhadani, Keith Currier, Al-Muthanna Al-Ani, Iryna Heiets, Tamara Oleshko, Dmytro Kvashuk, Anis Bousclet, Junsong Lei, Chaoqun Li, Zhaojiang Luo, Christopher Kitts, John Shepard, Alexander Ferrein, Micheal Reke, Stefan Schiffer, Ingrid Scholl, Gjorgji Nikolovski, Nicolas Limpert, Serdar Küçük

First published in London, United Kingdom, 2023 by IntechOpen
IntechOpen is the global imprint of INTECHOPEN LIMITED, registered in England and Wales, registration number: 11086078, 5 Princes Gate Court, London, SW7 2QJ, United Kingdom

British Library Cataloguing-in-Publication Data
A catalogue record for this book is available from the British Library

Additional hard and PDF copies can be obtained from orders@intechopen.com

Multi-Robot Systems - New Advances
Edited by Serdar Küçük
p. cm.

This title is part of the Artificial Intelligence Book Series, Volume 20
Topic: Multi-Agent System
Series Editor: Andries Engelbrecht
Topic Editor: Mehmet Emin Aydin

Print ISBN 978-1-83768-276-8
Online ISBN 978-1-83768-289-8
eBook (PDF) ISBN 978-1-83768-290-4
ISSN 2633-1403

For EU product safety concerns:
IN TECH d.o.o., Prolaz Marije Krucifikse Kozulić 3, 51000 Rijeka, Croatia, info@intechopen.com or visit our website at intechopen.com.

IntechOpen Book Series
Artificial Intelligence
Volume 20

Aims and Scope of the Series

Artificial Intelligence (AI) is a rapidly developing multidisciplinary research area that aims to solve increasingly complex problems. In today's highly integrated world, AI promises to become a robust and powerful means for obtaining solutions to previously unsolvable problems. This Series is intended for researchers and students alike interested in this fascinating field and its many applications.

Meet the Series Editor

Andries Engelbrecht received the Masters and Ph.D. degrees in Computer Science from the University of Stellenbosch, South Africa, in 1994 and 1999 respectively. He is currently appointed as the Voigt Chair in Data Science in the Department of Industrial Engineering, with a joint appointment as Professor in the Computer Science Division, Stellenbosch University. Prior to his appointment at Stellenbosch University, he has been at the University of Pretoria, Department of Computer Science (1998-2018), where he was appointed as South Africa Research Chair in Artifical Intelligence (2007-2018), the head of the Department of Computer Science (2008-2017), and Director of the Institute for Big Data and Data Science (2017-2018). In addition to a number of research articles, he has written two books, Computational Intelligence: An Introduction and Fundamentals of Computational Swarm Intelligence.

Meet the Volume Editor

Serdar Küçük received a BA and MSc from Marmara University, Istanbul, Turkey, in 1995 and 1998, respectively. He received a Ph.D. from Kocaeli University, Turkey, in 2004, where he is currently a full professor in the Department of Biomedical Engineering. He has several scientific publications to his credit, including international conference papers, journal papers, books, and book chapters. He serves as a reviewer of several well-known robotic journals. He is also an editor of scientific books. His research interests include optimization, control, and kinematics and dynamics modelling of serial and parallel robotic manipulators. Lately, Dr. Küçük has also been interested in designing electrical controlled, above-knee prosthetics and hand–wrist rehabilitation robots, surgical robots, and biomedical robotic devices.

Contents

Preface **XIII**

Section 1
Introduction **1**

Chapter 1 **3**
Introductory Chapter: Multi-Robot Systems Changing the Human Life
by Serdar Küçük

Section 2
Trajectory Planning for Robots **7**

Chapter 2 **9**
Trajectory Planning
by Junsong Lei, Zhaojiang Luo and Chaoqun Li

Section 3
Robot Control **25**

Chapter 3 **27**
A Multi-Layer, Multi-Robot Control Architecture for Long-Range, Dynamic Communication Links
by John Shepard and Christopher Kitts

Chapter 4 **53**
Controlling a Fleet of Autonomous LHD Vehicles in Mining Operation
by Alexander Ferrein, Gjorgji Nikolovski, Nicolas Limpert, Michael Reke, Stefan Schiffer and Ingrid Scholl

Chapter 5 **75**
Geometric Control of Robotic Systems
by Anis Bousclet

Section 4
Machine Learning and Vision **95**

Chapter 6 **97**
Challenges and Trends of Machine Learning in the Myoelectric Control System for Upper Limb Exoskeletons and Exosuits
by Jirui Fu, Zubadiah Al-Mashhadani, Keith Currier, Al-Muthanna Al-Ani and Joon-Hyuk Park

Chapter 7 **117**
The Use of Machine Vision in the Diagnosis of Ripening Strawberries
by Tamara Oleshko, Dmytro Kvashuk and Iryna Heiets

Preface

Interest in robotics has been steadily increasing. In recent years, multi-robot systems have started to attract the attention of students, academics, and industry workers. Recently, these systems have been used in many areas of industry, from load handling to simulators. Multi-robot systems are considered a useful alternative to conventional robot manipulators due to their better load-carrying capacity, high stiffness, better accuracy, and higher velocities and accelerations. However, the complexity of these systems presents some difficulties, such as end-effector positioning, control, and programming. Despite these difficulties, researchers are investigating multi-robot systems more and more every day.

Robotics technology is at a very interesting point, especially in terms of industrial robots, military robots, medical robots, rehabilitation robots, and micro robots. As a result of the hard work of academics and industrial robot designers, a new multi-robot design is developed almost daily.

This book explores the latest developments in multi-robot systems, with a focus on current robotic systems and applications. I would like to thank all the authors who contributed to this book with their valuable novel ideas and their knowledge of current developments in robotics.

Serdar Küçük, Ph.D.
Full Professor,
Technology Faculty,
Department of Biomedical Engineering,
University of Kocaeli,
Turkey

Section 1

Introduction

Chapter 1

Introductory Chapter: Multi-Robot Systems Changing the Human Life

Serdar Küçük

1. Introduction

People often do repetitive work in their daily lives, factories, schools, and hospitals. Most people do not like repetitive work and also do not like working in closed office environments. Especially, medical staff in hospitals perform thousands of repetitive tasks such as drawing blood and measuring the human heart rate. Factory workers also perform several never-ending repetitive tasks such as operating forklifts, painting cars, removing defective products, monitoring and troubleshooting, assembling products or components, operating machinery and tools, and packing products or components. People are faced with a much more important problem beyond being bored because of doing repetitive tasks. These repetitive jobs often cause work accidents such as poisoning from painting a car, injury from falling heavy metal in a factory, and hand cut from sharp knife. It is not an easy thing to make repairs under a ten-ton car for fear of falling on it. Even being in a hospital is a very stressful situation on its own. Working thousands of meters below in a coal mine is a frightening activity in itself. Working in a factory where explosives are produced is not for everyone. It is possible to multiply these examples.

At this very moment, multi-robot systems can find solutions to all these problems. Multi-robot systems are the best candidates for carrying out repetitive, monotonous, and dangerous tasks in the field without getting bored, tired, and overwhelmed. A medical robot can take blood from hundreds of patients without getting bored. A factory robot can nonstop paint cars for several days. A multi-robot system can tirelessly turn hundreds of cars into a line.

2. Multi-robot systems converting the human life

Almost eight decades have passed since the development of the first single serial robot manipulator. In the beginning, only serial robots were produced, but nowadays, multi-robot systems are designed and produced intensively. Multi-robot systems have recently started to be used safely in many areas. Today, multi-robot systems have become equipped enough to work alone in a factory, armed for use in battlefields, trusted to care for elderly people, skilled enough to participate in surgical operations in a hospital, excellent in pick-and-place applications, and smart enough to bring a selected product from a shopping mall. Especially as a result of the development and inevitable rise of multi-robot systems, it has become an inevitable reality that robots replace employees. Today, robots and human workers work in the same office. It seems inevitable that some people will leave their jobs entirely to robots in the future [1, 2]. In this case, social scientists should seek answers to the following questions: (1) How ready is humanity for this inevitable reality [3]? and (2) What solutions do social scientists propose for this inevitable reality [4]?

The fact that multi-robot systems are advanced enough to perform surgery in the hospital seems to benefit humanity as well as its bad consequences. Multi-robot systems are cheaper than designing all the skills in a single robot and doing the work. In particular, many robots can come together and do many things for the benefit of humanity. Swarm robotics is a very good example of this type of multi-robot system. Swarm robotics can be defined as the systematic management of a group consisting of a large number of simple robots. According to the some researchers, the collective behavior of robots in a robot swarm results from local interactions between robots [5]. This approach was inspired by other living things in nature, such as ants, where swarm behavior occurs. The field of swarm robotics studies the design, physical structures, and control behaviors of robots. There are several potential application areas of swarm robotics such as search and rescue missions [6] and military robot missions [7]. Autonomous robotic logistics is another multi-robot system that provides unmanned autonomous transfer of objects from one location to another with minimal human intervention. Surveillance and search and rescue applications are another application area of the multi robot systems. Surveillance practices are applied in both indoor [8] areas and outdoor areas [9]. Exploration is another application type of multi robot systems. In exploration, robots in a team are deployed to an unfamiliar location to explore the area. Many problems such as disconnection and battery drain are encountered in such applications [10].

3. Conclusion

The rapid improvement of technology causes robots to become more and more capable. Thus, robots can now perform tasks that they could not before, easier, faster, and with higher quality. In addition to these advantages, there are significant decreases in robot costs and operations compared to the past. As a result of falling robot costs, increasing skills, and working in cooperation, multi-robot systems have started to become an alternative for almost every job that humankind performs. Although multi-robot systems are increasingly working with people or alone in offices, the social consequences of this situation have not yet been fully explored. If people will work less in the future, how people will spend their free time is a fact that social scientists should think about in advance.

DOI: http://dx.doi.org/10.5772/intechopen.113168

Author details

Serdar Küçük
University of Kocaeli, Turkey

*Address all correspondence to: skucuk@kocaeli.edu.tr

References

[1] Küçük Ö. Computerization and automation: The declining importance of middle-skilled jobs and its impact on the workforce. In: Bulut M, Karacagil Z, editors. Sosyal Bilimlerde Güncel Tartışmalar 12. Bilgin Kültür Sanat Yayınları; 2023. Available from: https://www.bidgecongress.org/wp-content/uploads/2023/09/Sosyal-Bilimlerinde-Guncel-Tartismalar-12-17.pdf

[2] Küçük Ö. The effect of covid-19 crisis on a brain drain. In: Ağın K, editor. Sosyal Beşeri ve İdari Bilimler Alanında Uluslararası Araştırmalar IX. Eğitim Yayınevi; 2022. Available from: https://www.egitimyayinevi.com/u/egitimyayinevi/docs/y/o/yoksis-9786256382152-ekitap-sosyal-beseri-ix-icerik-dosyasi-1672407169.pdf

[3] Küçük Ö. Human capital as a new social policy subject with the World Bank. Gaziantep University Journal of Social Sciences. 2020;**19**(2):475-498. DOI: 10.21547/jss.6863865-2

[4] Küçük Ö. Beyond the impasse of market economy and technology, on discourse on the future of National States. In: Özkılbaç S, Bilgin T, editors. İktisat Alanında Seçilmiş Konular. Efe Akademi Yayınları; 2021. Available from: https://books.google.com.tr/books?id=yDJEEAAAQBAJ&printsec=frontcover&hl=tr&source=gbs_ge_summary_r&cad=0#v=onepage&q&f=false

[5] Dorigo M, Birattari M, Brambill M. Swarm robotics. Scholarpedia. 2014; **9**(1):1463. . DOI: 10.4249/scholarpedia.1463

[6] Hu J, Niu H, Carrasco J, Lennox B, Arvin F. Voronoi-based multi-robot autonomous exploration in unknown environments via deep reinforcement learning. IEEE Transactions on Vehicular Technology. 2020:14413-14423

[7] Madrigal Alexis C. Drone swarms are going to be terrifying and hard to stop. The Atlantic; 2018 [Accessed: March 7, 2019]

[8] Stancovici A, Micea MV, Cretu V. Cooperative positioning system for indoor surveillance applications. In: Proceedings of the 2016 International Conference on Indoor Positioning and Indoor Navigation (IPIN). Vol. 1. Madrid, Spain: IEEE; 4-7 Qct 2016. pp. 1-7

[9] Mendonça R, Marques MM, Marques F, Lourenço A, Pinto E, Santana P, et al. A cooperative multi-robot team for the surveillance of shipwreck survivors at sea. In: OCEANS 2016 MTS/IEEE Monterey [7761074]. Institute of Electrical and Electronics Engineers (IEEE); 2016. DOI: 10.1109/OCEANS.2016.7761074

[10] Banfi J, Li AQ, Basilico N, Rekleitis I, Amigoni F. Asynchronous multirobot exploration under recurrent connectivity constraints. In: IEEE International Conference on Robotics and Automation. Stockholm, Sweden: IEEE; 2016. pp. 5491-5498

Section 2

Trajectory Planning for Robots

Chapter 2

Trajectory Planning

Junsong Lei, Zhaojiang Luo and Chaoqun Li

Abstract

This chapter discusses the trajectory planning of robot. First, the basic principle of trajectory planning is described, which is mainly determined by the geometric path and the motion law, and it is a motion law that defines time according to a given geometric path. Second, the characteristics of the joint space and the operational space are expounded. Then the trapezoidal trajectory and the trajectory with S velocity profile commonly used in industrial practice are introduced. Finally, point-to-point trajectories and multipoint trajectories are described. Point-to-point trajectories include a point-to-point trajectory in the joint space and a straight-line and circle trajectory in the operational space. Multipoint trajectories include Bezier and NURBS curves functions.

Keywords: trajectory planning, the geometric path, the motion law, the joint space, the operational space, point-to-point trajectories, multipoint trajectories

1. Introduction

Trajectory planning is a motion law that defines time according to a given geometric path. Therefore, the purpose of trajectory planning is to meet the needs of the tasks and generate the reference inputs required by the control system, so that the robots can perform the movements in Ref. [1]. The inputs of any trajectory planning algorithm are an expected motion path, kinematics, and dynamics system parameters of the robots, and the outputs are the positions, velocities, and accelerations of the interpolation points of each joint or the end effector in Refs. [2, 3]. This chapter will introduce trajectory planning from the perspective of industrial practice, including the basic principle of trajectory planning, the joint space and the operational space, the motion laws, and point-to-point and multipoint trajectories.

2. The basic principle of trajectory planning

The desired trajectory can be fully determined by the geometric path and the motion law. The geometry path is related to space, while the motion law is related to time.

2.1 The geometry of the trajectory

The geometric path $p = p(s)$ expresses the path of the end effector from the start point to the end point in Ref. [4]. The end point B can be reached in two different path

IntechOpen

forms from the start point A, see **Figure 1**. Depending on whether it is necessary to specify the geometry of the trajectory, the planning of trajectory can be carried out in the joint space or the operational space.

2.2 The motion law

The motion law $s = s(t)$ limits velocity, acceleration, and other parameters of the end effector from the start point to the end point. The end point B can be reached from the start point A according to two different motion laws, see **Figure 2**. Various functions, such as linear trajectory function with constant velocity or trajectory function with S velocity profile, can be selected to specify the motion law along the geometric path.

Therefore, based on the geometric path $p = p(s)$ and the motion law $s = s(t)$, the trajectory can be written as

$$p = p(s(t)) \tag{1}$$

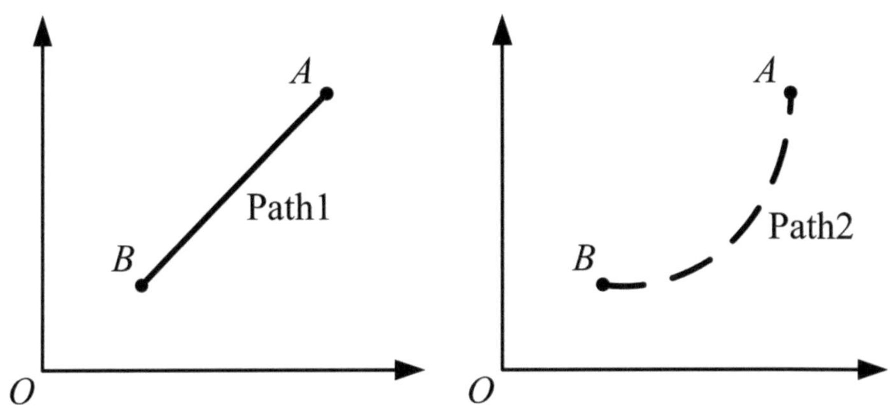

Figure 1.
Two different path forms from the start point A to the end point B.

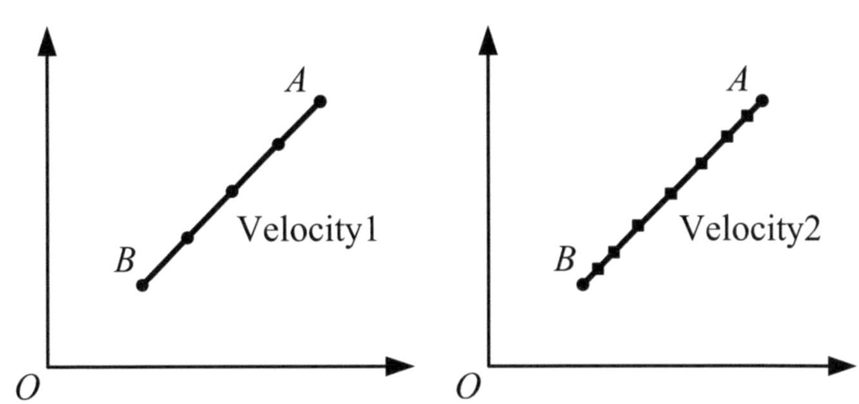

Figure 2.
Two different motion laws from the start point A to the end point B.

3. The joint space and the operational space

Trajectory planning is divided into the joint space and the operational space according to the different description spaces. Trajectory planning in the joint space has the characteristics of few constraints and fast calculation speed, while trajectory planning in the operational space is mainly used for scenarios where users have specified path requirements.

3.1 The joint space

Trajectory planning in the joint space is to describe the trajectory of the robot as a function of joint position with respect to time, that is, the trajectory of the end effector can be directly determined by joint variables, so it is easy and simple to carry out trajectory planning in the joint space. Since there is no continuous corresponding relationship between the joint space and the operational space, singularity, and redundant degrees of freedom in motion can be avoided. In addition, the complexity of its path depends on the specific kinematic characteristics of the robot, and the path in the operational space cannot be predicted in Ref. [5]. Therefore, trajectory planning in the joint space is only applicable to the operating scenarios without path requirements.

3.2 The operational space

Trajectory planning in the operational space is based on a function of position and orientation with respect to time to describe the trajectory of the robot, and the corresponding joint position is obtained by the inverse kinematic model.

Since the path planned in the operational space is determined, the trajectory planning in the operational space is suitable for the operating scenarios with strict requirements for the instantaneous change law of position and orientation. In addition, the planning of trajectory in the operational space is prone to problems such as unreachable workspace or singularity, and they are generally avoided by specifying the robots to pass through a set of intermediate points in Ref. [6].

4. The motion law

In order to ensure that the robots can operate efficiently and stably, it is generally required that the motion profile meet at least the continuous first derivative, and it needs to have the characteristics of fast acceleration and deceleration and constant velocity. At present, the velocity control algorithms commonly used in robot systems include the trapezoidal trajectory and the trajectory with S velocity profile.

4.1 The trapezoidal trajectory

The trapezoidal trajectory means that the velocity profile is trapezoidal, and its complete curves are shown in **Figure 3**. The complete trajectory consists of three phases. In the first phase, the acceleration changes from zero to a fixed acceleration,

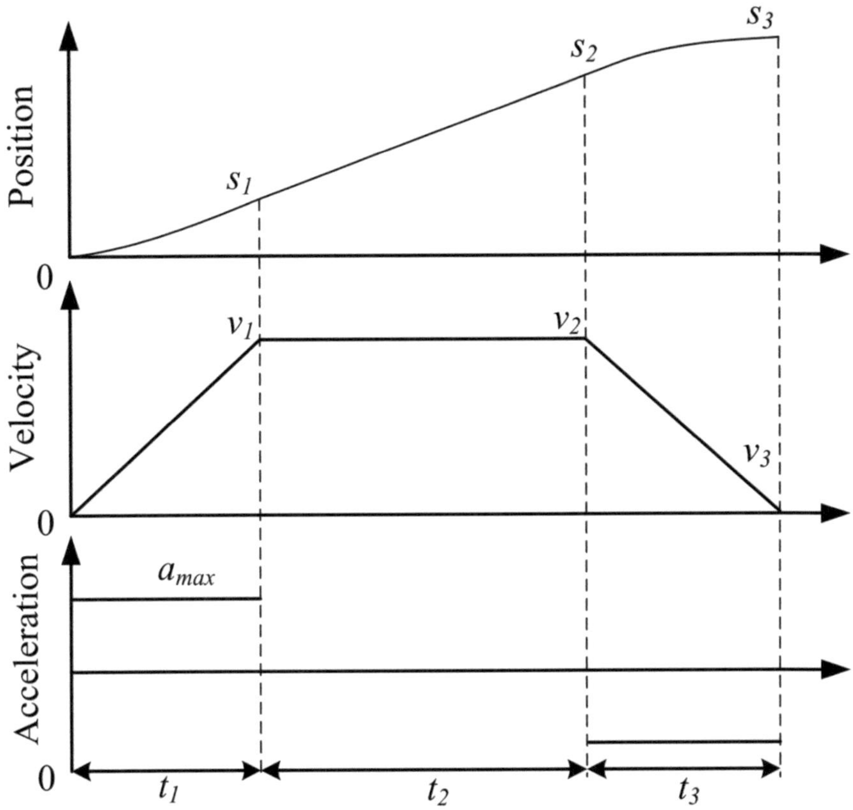

Figure 3.
Position, velocity, and acceleration curves of a trapezoidal trajectory.

and the velocity increases linearly with a fixed acceleration from zero. In the second phase, when the velocity reaches the desired value, it enters the constant velocity phase. At this phase, the velocity is constant and the acceleration suddenly changes to zero. In the third phase, the velocity begins to decrease with a fixed acceleration until it reaches zero.

Piecewise acceleration equations are expressed as

$$a(t) = \begin{cases} a_{\max} & (0 \leq t \leq t_1) \\ 0 & (t_1 \leq t \leq t_2) \\ -a_{\max} & (t_2 \leq t \leq t_3) \end{cases} \tag{2}$$

Piecewise velocity equations are expressed as

$$v(t) = \begin{cases} a_{\max} t & (0 \leq t \leq t_1) \\ v_2 & (t_1 \leq t \leq t_2) \\ v_2 - a_{\max}(t - t_2) & (t_2 \leq t \leq t_3) \end{cases} \tag{3}$$

Piecewise position equations are expressed as

$$s(t) = \begin{cases} \frac{1}{2} a_{max} t^2 & (0 \le t \le t_1) \\ s_1 + v_2(t - t_1) & (t_1 \le t \le t_2) \\ s_2 - \frac{1}{2} a_{max}(t - t_2)^2 & (t_2 \le t \le t_3) \end{cases} \tag{4}$$

where the time instants t_1, t_2 and t_3 are specified. s_1, s_2 and s_3 are corresponding position values, and v_1, v_2 and v_3 are corresponding velocity values. a_{max} is a maximum acceleration value.

4.2 The trajectory with S velocity profile

The trajectory with S velocity profile means that the velocity profile is S-shaped, which presents that the jerk is constant during acceleration and deceleration, and a smoother velocity curve is generated by controlling the duration of jerk to reduce the impact on machinery. The complete trajectory with S velocity profile includes seven phases: constant positive jerk phase T_1 in acceleration, constant zero jerk phase T_2 in acceleration, constant negative jerk phase T_3 in acceleration, constant velocity phase T_4, constant negative jerk phase T_5 in deceleration, constant zero jerk phase T_6 in deceleration, and constant positive jerk phase T_7 in deceleration, see **Figure 4**. In the first three phases, the velocity increases smoothly, the acceleration presents trapezoidal profile, and the jerk changes abruptly between the desired value and zero. In the fourth phase, when the velocity is accelerated to the desired value, it enters the constant velocity phase. When the end point is approaching, the deceleration phase begins in the last phases. The transition of the entire velocity curve is smooth, and there is a continuous acceleration.

Piecewise jerk equations are expressed as

$$j(t) = \begin{cases} j_{max} & (0 \le t \le t_1) \\ 0 & (t_1 \le t \le t_2) \\ -j_{max} & (t_2 \le t \le t_3) \\ 0 & (t_3 \le t \le t_4) \\ -j_{max} & (t_4 \le t \le t_5) \\ 0 & (t_5 \le t \le t_6) \\ j_{max} & (t_6 \le t \le t_7) \end{cases} \tag{5}$$

Piecewise acceleration equations are expressed as

$$a(t) = \begin{cases} j_{max} t & (0 \le t \le t_1) \\ j_{max} T_1 & (t_1 \le t \le t_2) \\ j_{max} T_1 - j_{max}(t - t_1) & (t_2 \le t \le t_3) \\ 0 & (t_3 \le t \le t_4) \\ -j_{max}(t - t_4) & (t_4 \le t \le t_5) \\ -j_{max} T_4 & (t_5 \le t \le t_6) \\ -j_{max} T_4 + j_{max}(t - t_6) & (t_6 \le t \le t_7) \end{cases} \tag{6}$$

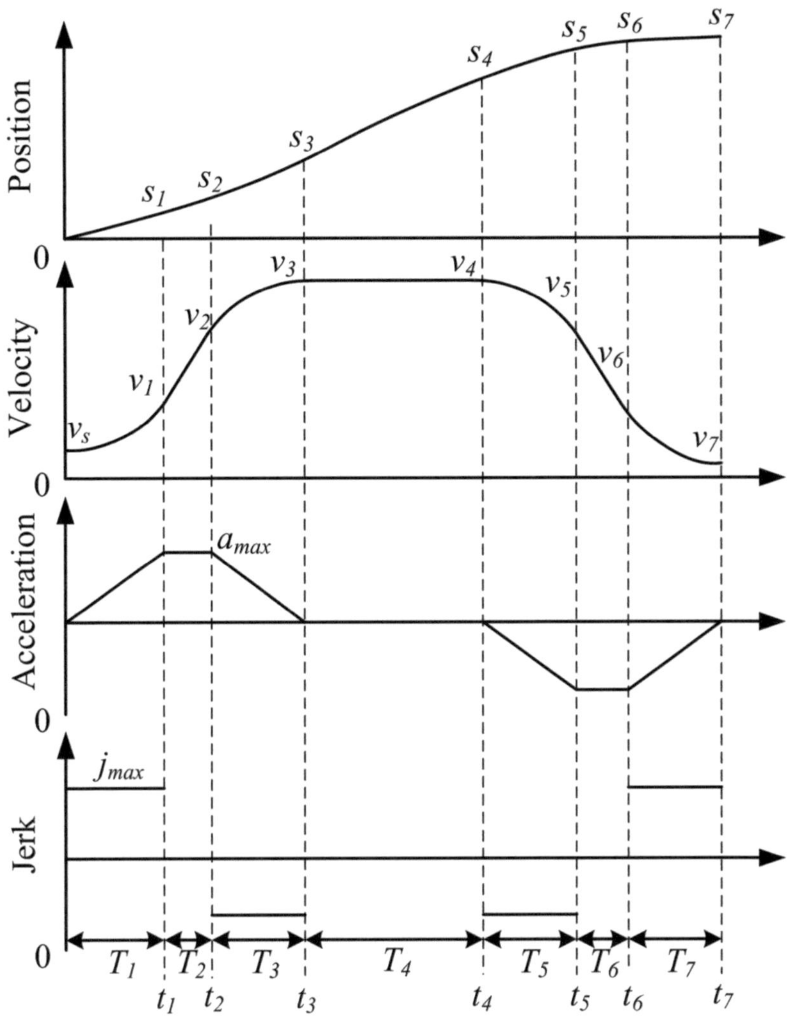

Figure 4.
Position, velocity, and acceleration curves of a trajectory with S velocity profile.

Piecewise velocity equations are expressed as

$$v(t) = \begin{cases} v_s + \frac{1}{2} j_{\max} t^2 & (0 \le t \le t_1) \\ v_1 + j_{\max} T_1 (t - t_1) & (t_1 \le t \le t_2) \\ v_2 + j_{\max} T_1 (t - t_2) - \frac{1}{2} j_{\max} (t - t_2)^2 & (t_2 \le t \le t_3) \\ v_3 & (t_3 \le t \le t_4) \\ v_4 - \frac{1}{2} j_{\max} (t - t_4)^2 & (t_4 \le t \le t_5) \\ v_5 - j_{\max} T_5 (t - t_5) & (t_5 \le t \le t_6) \\ v_6 - j_{\max} T_5 (t - t_6) + \frac{1}{2} j_{\max} (t - t_6)^2 & (t_6 \le t \le t_7) \end{cases} \tag{7}$$

Piecewise position equations are expressed as

$$s(t) = \begin{cases} v_s t + \frac{1}{6} j_{\max} t^3 & (0 \le t \le t_1) \\ s_1 + v_1(t - t_1) + \frac{1}{2} j_{\max} T_1 (t - t_1)^2 & (t_1 \le t \le t_2) \\ s_2 + v_2(t - t_2) + \frac{1}{2} j_{\max} T_1 (t - t_2)^2 - \frac{1}{6} j_{\max} (t - t_2)^3 & (t_2 \le t \le t_3) \\ s_3 + v_3(t - t_3) & (t_3 \le t \le t_4) \\ s_4 + v_4(t - t_4) - \frac{1}{6} j_{\max} (t - t_4)^3 & (t_4 \le t \le t_5) \\ s_5 + v_5(t - t_5) - \frac{1}{2} j_{\max} T_5 (t - t_5)^2 & (t_5 \le t \le t_6) \\ s_6 + v_6(t - t_6) - \frac{1}{2} j_{\max} T_5 (t - t_6)^2 + \frac{1}{6} j_{\max} (t - t_6)^3 & (t_6 \le t \le t_7) \end{cases} \tag{8}$$

Where the time instants $t_1, t_2, \dots,$ and t_7 are specified. $s_1, s_2, \dots,$ and s_7 are corresponding position values, and $v_1, v_2, \dots,$ and v_7 are corresponding velocity values. $a_{\max}$ and $j_{\max}$ are maximum acceleration and jerk values.

5. Point-to-point trajectories and multipoint trajectories

5.1 Point-to-point trajectories

Point-to-point trajectories are mainly used for robotic tasks such as palletizing, cutting, press tending, and other operations. Point-to-point control first ensures the position and orientation of the teach-in points, and the geometric path between the teach-in points, such as straight-lines and circles, can be specified when programming the robot's working program.

5.1.1 Point-to-point trajectory in the joint space

According to the requirements of the task, the end effector is required to move linearly in the joint space between point A and point B (**Figure 5**).

Therefore, the position of each joint at the time instant $t + \Delta t$ can be expressed as

$$\theta_{t+\Delta t} = (1 - u)\theta_A + u\theta_B (0 \le u \le 1) \tag{9}$$

5.1.2 Point-to-point trajectory in the operational space

5.1.2.1 A straight-line position trajectory

According to the requirements of the task, the position of the end effector needs to be moved in a straight-line between point A and point B. Generally, the positions of the start point A and the end point B in the operational space are known, and the positions of a set of intermediate points on the trajectory AB are calculated, see **Figure 6**.

The position length of the trajectory AB is L_{AB}. The end effector is located at the point P_t and $P_{t+\Delta t}$ on the trajectory AB, corresponding to the displacement $s(t)$ and

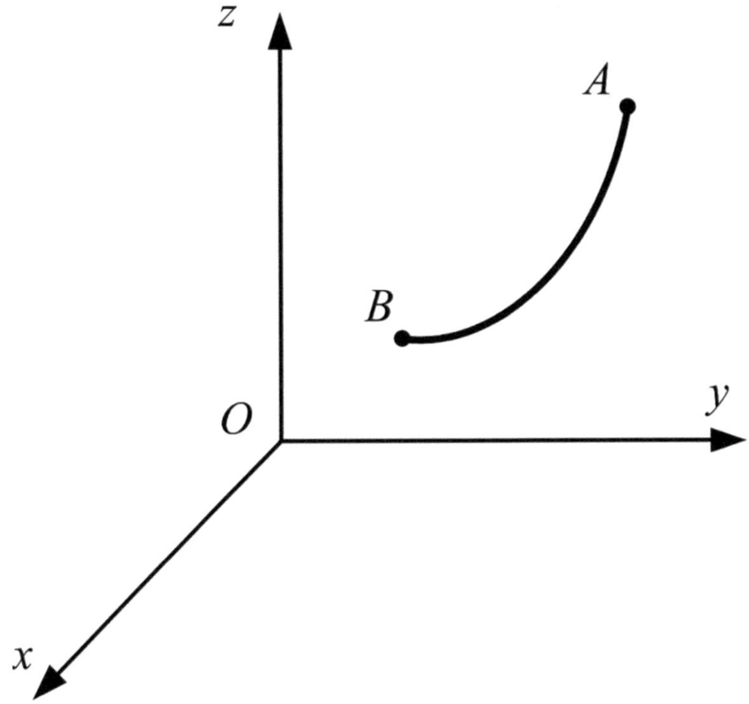

Figure 5.
A point-to-point trajectory in the joint space.

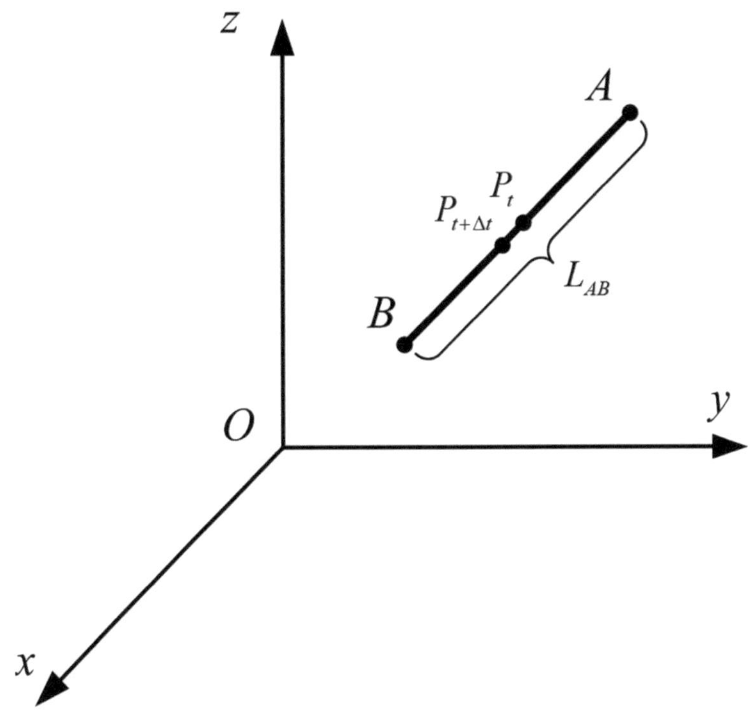

Figure 6
A straight-line position trajectory in the operational space.

$s(t+\Delta t) = s(t) + \Delta s$ at the time instants t and $t+\Delta t$. Δt is a control period, corresponding to the displacement Δs .

Length function in normalized form can be expressed as

$$u = \frac{s(t+\Delta t)}{L_{AB}} (0 \leq u \leq 1) \tag{10}$$

Therefore, the position of the end effector at the time instant $t+\Delta t$ can be expressed as

$$\begin{bmatrix} P_x \\ P_y \\ P_z \end{bmatrix} = (1-u) \begin{bmatrix} A_x \\ A_y \\ A_z \end{bmatrix} + u \begin{bmatrix} B_x \\ B_y \\ B_z \end{bmatrix} \tag{11}$$

5.1.2.2 A straight-line orientation trajectory

According to the requirements of the task, the orientation of the end effector needs to be moved continuously between point A and point B. Generally, the orientations of the start point A and the end point B in the operational space are known, and the orientations of a set of intermediate points on the trajectory AB are calculated, see **Figure 7**. Since Euler angles and rotation matrices may have gimbal locking in rotation, and quaternions do not have this problem. Therefore, quaternions can be used to calculate the orientation.

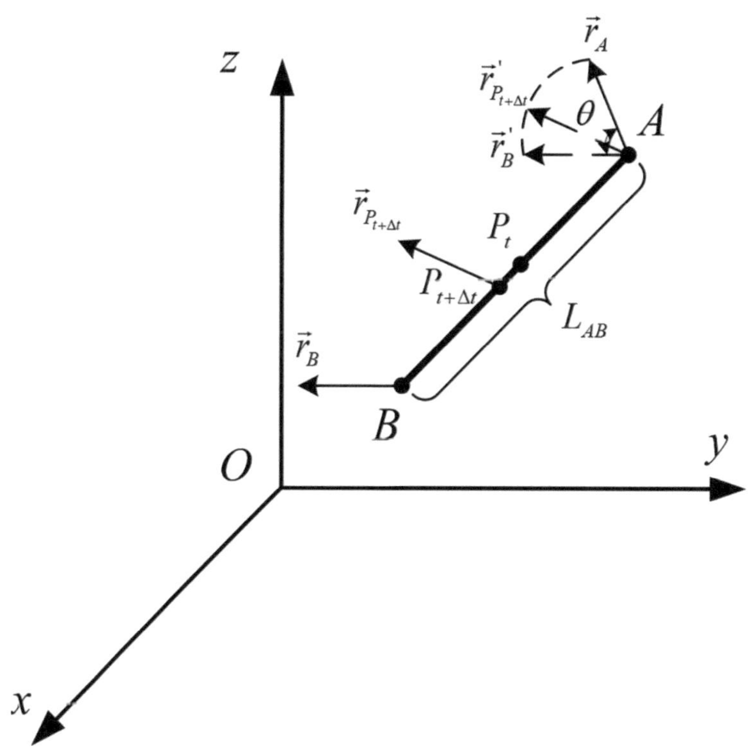

Figure 7
A straight-line orientation trajectory in the operational space.

The orientation unit vector of the start point A is $\vec{r}_A$, corresponding to the quaternion Q_A. The orientation unit vector of the end point B is $\vec{r}_B$, corresponding to the quaternion Q_B. The angle between two orientation unit vectors is θ. The end effector is located at the point P_t and $P_{t+\Delta t}$ on the trajectory AB, corresponding to the angle $\theta(t)$ and $\theta(t+\Delta t)$ at the time instants t and $t+\Delta t$. Its orientation unit vector is $\vec{r}_{P_{t+\Delta t}}$, and the corresponding quaternion is $Q_{P_{t+\Delta t}}$.

Angle function in normalized form can be expressed as

$$u = \frac{\theta(t+\Delta t)}{\theta}(0 \le u \le 1) \tag{12}$$

The general linear interpolation will cause uneven changes in angular velocity, while spherical linear interpolation can ensure a smooth curve between two quaternions. Therefore, the orientation of the end effector at the time instant $t+\Delta t$ can be expressed as

$$Q_{P_{t+\Delta t}} = \frac{\sin[(1-u)\theta]Q_A + \sin(u\theta)Q_B}{\sin\theta} \tag{13}$$

When the angle θ is very small, it will result in a large value in the Eq. (13). Therefore, the linear interpolation formula can be used

$$Q_{P_{t+\Delta t}} = (1-u)Q_A + uQ_B \tag{14}$$

5.1.2.3 A circle position trajectory

According to the requirements of the task, the position of the end effector needs to be moved in a circle between points A, B, and C. Generally, the positions of the start point A, the middle point B and the end point C in the operational space are known,

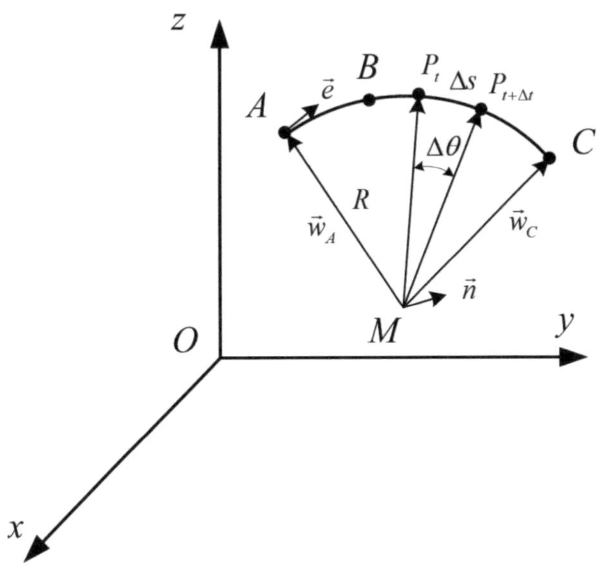

Figure 8
A circle position trajectory in the operational space.

and the positions of a set of intermediate points on the trajectory ABC are calculated, see **Figure 8**. Based on the position of the start point A, the middle point B, and the end point C, the radius R and center M of the circle can be solved. Then the unit normal vector $\vec{n}$ of the circle, the unit tangent vector $\vec{e}$ of the start point, and the unit vector $\vec{w}_A$ from the center to the start point can be solved.

Length function in normalized form can be expressed as

$$u = \frac{s(t + \Delta t)}{L_{\widehat{ABC}}} (0 \le u \le 1) \tag{15}$$

Therefore, the position of the end effector at the time instant $t + \Delta t$ can be expressed as

$$\begin{bmatrix} P_x \\ P_y \\ P_z \end{bmatrix} = \begin{bmatrix} M_x \\ M_y \\ M_z \end{bmatrix} + R\bullet \cos\left(\frac{uS}{R}\right)\vec{w}_A + R\bullet \sin\left(\frac{uS}{R}\right)\vec{e} \tag{16}$$

5.1.2.4 A circle orientation trajectory

According to the requirements of the task, the orientation of the end effector needs to be moved continuously between points A, B, and C. Generally, the orientations of the start point A, the middle point B and the end point C in the operational space are known, and the orientations of a set of intermediate points on the trajectory ABC are calculated, see **Figure 9**.

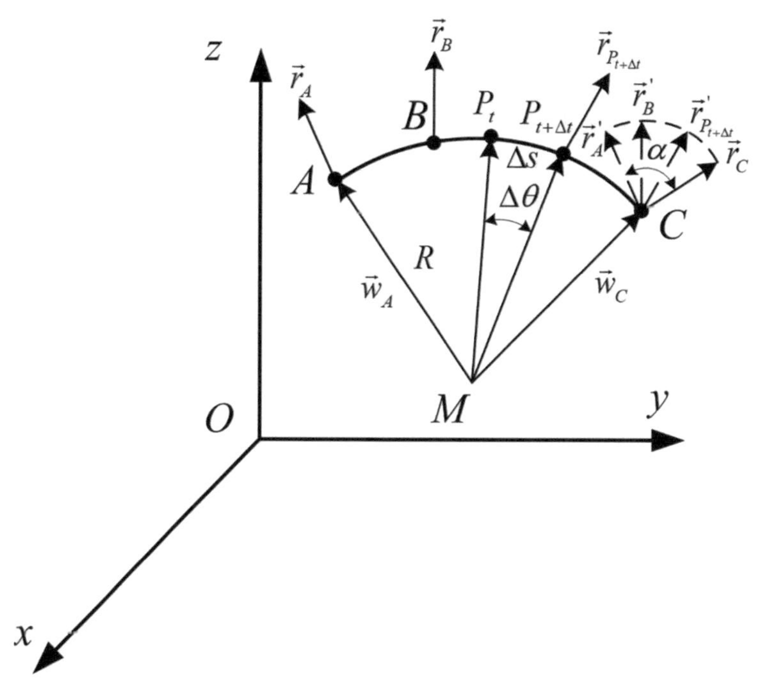

Figure 9.
A circle orientation trajectory in the operational space.

The orientation unit vector of the start point A is $\vec{r}_A$, corresponding to the quaternion Q_A. The orientation unit vector of the middle point B is $\vec{r}_B$, corresponding to the quaternion Q_B. The orientation unit vector of the end point C is $\vec{r}_C$, corresponding to the quaternion Q_C. The angle of the orientation unit vectors between the start and the middle point is α_{AB}. The angle of the orientation unit vectors between the middle and end point is α_{BC}. The angle of the orientation unit vectors between the start and end point is α. The end effector is located at the point $P_{t+\Delta t}$ on the trajectory AB, corresponding to the angle $\alpha(t + \Delta t)$ at the time instant $t + \Delta t$. Its orientation unit vector is $\vec{r}_{P_{t+\Delta t}}$, and the corresponding quaternion is $Q_{P_{t+\Delta t}}$.

Middle angle function in normalized form can be expressed as

$$u_{Middle} = \frac{\alpha_{AB}}{\alpha} \tag{17}$$

Angle function in normalized form can be expressed as

$$u = \frac{\alpha_{t+\Delta t}}{\alpha} \tag{18}$$

If $u \le u_{Middle}$, the orientation of the end effector at the time instant $t + \Delta t$ can be expressed as

$$Q_{P_{t+\Delta t}} = \frac{\sin\left[\left(1 - \frac{u}{u_{Middle}}\right)\alpha_{AB}\right]Q_A + \sin\left(\frac{u}{u_{Middle}}\alpha_{AB}\right)Q_B}{\sin\alpha_{AB}} \tag{19}$$

If $u_{Middle} \le u$, the orientation of the end effector at the time instant $t + \Delta t$ can be expressed as

$$Q_{P_{t+\Delta t}} = \frac{\sin\left[\left(1 - \frac{u-u_{Middle}}{1-u_{Middle}}\right)\alpha_{BC}\right]Q_B + \sin\left(\frac{u-u_{Middle}}{1-u_{Middle}}\alpha_{BC}\right)Q_C}{\sin\alpha_{BC}} \tag{20}$$

5.2 Multipoint trajectories

Multipoint trajectories are often used for complex operations such as painting, arc welding, or surface machining. The user teaches a series of points (such as start, end, and intermediate points) on a continuous path, and the robots connect the path points or multi-segment paths sequentially using polynomial functions, Bezier curve functions, or spline functions.

5.2.1 Bezier curve functions

In order to further improve the efficiency of the tasks, the robot end effector is required to achieve continuous operation between multiple trajectories. That is to say, the displacement and speed motion profiles are continuous. Due to the characteristics of the geometric invariance and convex hull, high-order Bezier curves are often used in industrial practice to connect multi-segment paths in Ref. [7]. Taking the connection of two straight-line trajectories as an example, the position of point A, point B and point C of line 1 and line 2 is generally known, and the end point B of line 1 is used as the start point of line 2, see **Figure 10**.

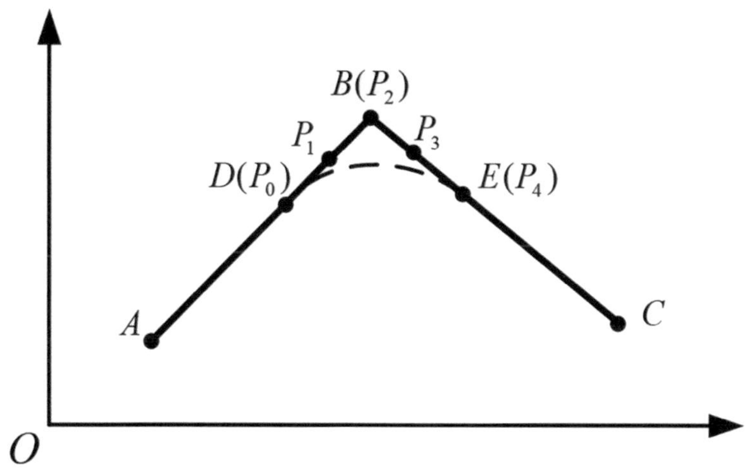

Figure 10.
A quartic Bezier curve trajectory.

A circle with B as center point and the given transition value R as radius can be obtained, and it intersects the AB and BC trajectories at D and E points, respectively. A quartic Bezier curve is used to connect the two trajectories, and five control points need to be specified on the DB and BE segments. The center point B is the control point P_2, and the points D and E are the control points P_0 and P_4, respectively. P_1 is taken halfway through the DB segment and P_3 is taken halfway through the BE segment.

The quartic Bezier curve function can be written as

$$P(t) = \sum_{i=0}^{4} P_i B_{i,4}(t), t \in [0, 1] \tag{21}$$

The corresponding expansion in Eq. (21) is

$$\begin{aligned} P(t) = (1-t)^4 P_0 + 4t(1-t)^3 P_1 + 6t^2(1-t)^2 P_2 \\ + 4t^3(1-t) P_3 + t^4 P_4, t \in [0, 1] \end{aligned} \tag{22}$$

Recursive equation is

$$P_i^k = \begin{cases} P_i & k = 0 \\ (1-t)P_i^{k-1} + tP_{i+1}^{k-1} & k = 1, 2, \ldots, 4, i = 0, 1, \ldots, 4-k \end{cases} \tag{23}$$

5.2.2 NURBS curve functions

For the complex trajectories of the end effector in the operation space, the NURBS curve has been widely used because it can accurately and uniformly express standard analytical curves and free curves and has the characteristic of flexible shape control in Ref. [8]. Any k-th degree NURBS curve can be represented as a piecewise rational polynomial function.

$$P(t) = \frac{\sum_{i=0}^{n} d_i \omega_i N_{i,k}(t)}{\sum_{i=0}^{n} \omega_i N_{i,k}(t)} \tag{24}$$

where ω_i are the weight factors, d_i are the curve control points, and $N_{i,k}(t)$ are B-spline basis functions determined by the node vectors t_i .

Based on De Boor-Cox recursive function, the basis function can be written as

$$\begin{cases} N_{i,0} = \begin{cases} 1 \ , t \in [t_i, t_{i+1}) \\ 0 \ \ , otherwise \end{cases} \\ N_{i,k} = \dfrac{t - t_i}{t_{i+k} - t_i} N_{i,k-1}(t) + \dfrac{t_{i+k+1} - t}{t_{i+k+1} - t_{i+1}} N_{i+1,k-1}(t) \end{cases} \tag{25}$$

where $\frac{0}{0} = 0$ is specified in Eq. (25).

Due to the computational complexity of the control points solved by the via-points of the NURBS curve, the De Boor-Cox algorithm is often used to avoid repeated iteration. More details on NURBS curve calculations can be found in Ref. [9].

6. Conclusions

This chapter discusses the trajectory planning of robots. First, the basic principle of trajectory planning is described, which is mainly determined by the geometric path and the motion law. Second, the characteristics of the joint space and the operational space are expounded. Then the trapezoidal trajectory and the trajectory with S velocity profile commonly used in industrial practice are introduced. Finally, point-to-point trajectories and multipoint trajectories are described. Point-to-point trajectories include a point-to-point trajectory in the joint space and a straight-line and circle trajectory in the operational space. Multipoint trajectories include Bezier and NURBS curves functions.

Author details

Junsong Lei[1,2]*, Zhaojiang Luo[2] and Chaoqun Li[2]

1 Key Laboratory of High Performance Servo Systems in Guangdong Province, China

2 Zhuhai Gree Electric Co., Ltd., Zhuhai, China

*Address all correspondence to: rey_se@163.com

References

[1] Zixing C, Bin X. Robotics. 3rd ed. Beijing: Tsinghua Press; 2015. p. 273

[2] Caselli S, Reggiani M, Sbravati R. Parallel path planning with multiple evasion strategies. In: Proceedings of the IEEE International Conference on Robotics and Automation (ICRA '02); 11–15 May 2002; Washington, DC. New York: IEEE; 2002. pp. 260-266

[3] Yonglun C, Shiqiang Z, Lijia L. Kinematics analysis and simulation of QJ-6R welding robot based on Matlab. Mechanical and Electrical Engineering Magazine. 2007;**24**:107-110. DOI: 10.3969/j.issn.1001-4551.2007.11.034

[4] Luigi B, Claudio M. Trajectory Planning for Automatic Machines and Robots. Berlin: Springer-Verlag; 2008. pp. 7-9

[5] John JC. Introduction to Robotics: Mechanics and Control. 3rd ed. Beijing: China Machine Press; 2012. pp. 161-179

[6] Saeed NB. Introduction to Robotics: Analysis, Control, Applications. 2nd ed. Beijing: Electronic Industry Press; 2013. pp. 136-153

[7] Alba-gomez OG, Pamanes JA, Wenger P. Trajectory planning of parallel manipulators for global performance optimization. In: International Symposium on Advances in Robot Kinematics, 1-5 January 2008. Netherlands: Springer; 2008. pp. 253-261

[8] Faires JD, Burden RL. Numerical Methods. 4th ed. Pacific Grove: Brooks/Cole Publishing Company; 2012. pp. 157-170

[9] Les P, Wayne T. The NURBS Book. 2nd ed. Beijing: Tsinghua Press; 2010. pp. 32-72

Section 3

Robot Control

Chapter 3

A Multi-Layer, Multi-Robot Control Architecture for Long-Range, Dynamic Communication Links

John Shepard and Christopher Kitts

Abstract

A unified motion control architecture is presented for dynamic, long-range multi-robot communications networks, incorporating task abstraction that disassociates goals from implementation. In the task space, communication link states are specified, directly measured, and explicitly controlled yielding well-behaved task state trajectories. The control architecture uses task-level compensation to generate multi-robot formation mobility commands, and a cluster space controller transforms those formation commands to mobility commands for individual robots. The number of robots are selected to meet communications requirements and controlled through a multi-task coordination capability incorporated within the architecture. Robustness to performance commands, system configuration parameters, and external disturbances is demonstrated through a variety of simulations and experiments. These show how robots are dynamically positioned and switched into or out of operation in order to meet communications requirements.

Keywords: multi-robot systems, mobile robots, communication networks, robot control, collaborative robots

1. Introduction

Robotic systems are an integral tool in modern society, extending the capability of human operators, increasing their productivity and improving their quality of life. Multi-robot systems are able to enhance quantity-sensitive performance metrics like speed, coverage, throughput, and redundancy. In addition, their spatial diversity provides added capabilities for tasks like formation-keeping [1–3], escorting or guarding [4, 5], surveillance and feature tracking [6–9], object manipulation [10], and more specialized tasks such as automatic lighting [11] reconfigurable sparse antenna arrays [12], and minimally invasive surgery [13, 14]. New research is developing techniques for diverse task-oriented groups of robots to work together to perform broader and more sophisticated missions [15, 16]; as the mission evolves, the tasks may be preserved even though the environment, the performance objectives, and the assignment of robots may change.

In this research, we examine the task of long-range and dynamic communication link management due to its necessity as a supporting role in many applications and missions. By "long-range" we refer to communication links that require a single series of repeater stations to relay communication between remote end stations. As such, mesh network approaches such as those described in [17, 18] which provide multiple communication paths are not appropriate for the applications of interest. By "dynamic" we refer to the need for robot relays to adjust their positions to compensate for changing link conditions due to motion of the end stations, the attenuation environment, communications equipment performance, etc.; as such, deploying static relay stations, as is done in [19, 20], is not sufficient.

Model-based approaches to link management use a model of the link along with information regarding relay robot positions in order to estimate quality of service. These are typically well behaved due to smoothing simplifications [21] but can be inaccurate. In [22], a simple binary model is used, where nodes are assumed to be connected if within a fixed distance, leading to a strategy that evenly spaces robots between end-stations. More sophisticated models may be used, incorporating path models based on power, distance, and line-of- sight in order to identify a set of goal positions for the robots [23–25].

Measurement-based approaches remove model inaccuracies, allowing for improved performance and an expanded workspace [26], and are robust to complexities of the link behavior like obstructions, directional antenna patterns, and multipath effects. For example, in [27], unmanned aircraft are used to establish a relay network, with the airborne relays circling control points and measuring link gradients to relocate to optimal locations. In [28], it was found that the use of measurement-based approaches improved performance in maintaining a line of sight connection between robots. In [29], optimization techniques exploited measurements to minimize the path traveled by relay robots as they manuever to a fixed location to support end-station communication. In [30], a centralized planner guides a multirobot team in known indoor environments to establish a multihop network, with measurements used to improve local positioning. Finally, in [31, 32], multi-robot mesh communications is achieved using a potential-function-based, decentralized control scheme and measurements of communication bit error rate.

The work presented in this paper addresses repositioning of communication relay nodes operating within a long-range, dynamic link in order to ensure that a measured Quality of Service (QoS) level, signal strength, is maintained. Service is achieved and maintained using a multi-layer control architecture. At the highest level, a task-oriented operational space control approach produces well-behaved task state compensations. These are converted to robot formation motion commands using a model-based inverse Jacobian transform. Formation compensations are then converted to individual robot drive commands to achieve the necessary level of position control. A flexible number of robots is used to support the link, with robots being switched into and out of service as necessary; to be more precise, unused robots are actually switched to a benign position-control task that maintains their location in a ready position for future use in maintaining the link. Formally, all robots, regardless of their assigned task, are controlled through a single unified position control framework that dynamically changes the transforms used to convert between control state spaces.

Each element of this control architecture is well-motivated and provides a useful innovation compared to previous work in managing dynamic long-range communication links. First, direct sensing of and control implementation in the operational task space improves application-oriented performance. Second, use of this space abstracts the specification of desired performance from implementation details such as the number of robots used to achieve the task and the mobility characteristics of these robots, thereby preventing the redesign of the specification process when these implementation details are changed. Third, our approach flexibly engages an appropriate number of robots in the link management task, thereby conserving resources when the full suite of robots is not required. Fourth, use of the intermediate cluster space representation provides a critical layer of abstraction that reflects the geometric nature of the application, thereby making the construction of task level transforms simpler; furthermore, this control architecture provides an intuitive intermediate layer for supervisory operators when specifying and monitoring performance as well as for developers when incrementally designing, verifying and troubleshooting functionality of the multi-robot system. Fifth, the architecture unifies motion control for all robots within a single control architecture (whether the controller itself is centralized or decentralized) thereby facilitating development and performance analysis of large-scale, multi-task missions. Overall, our approach provides enhanced performance, minimizes the use of robot resources, promotes modular composition, and has been verified experimentally; results show the technique to be robust to the dynamic link environment based on moving end stations, local attenuation, and variation in the state of the communication relays.

The article starts with a synopsis of the basic control architecture in Section 2, discusses extension to the task space in Section 3, integrates multiple tasks in Section 4, describes the experimental test bed in Section 5, presents results of the experiments and simulations in Sections 6, and considers future work in Section 7.

2. Cluster space control

The first layer of our control architecture addresses mobility control of the multi-robot cluster. While a number of techniques exist for multi-robot formation control, the Cluster Space control architecture abstracts the cluster as a virtual articulating mechanism, allowing explicit control of all system states. The interested reader should consult [33] for comparisons. The underlying goal of the Cluster Space technique is simple motion specification and control of multi-robot systems. This is accomplished by considering multiple robots as a single geometric entity. The pose of a cluster is described by its location and shape, which are related to individual robot positions through a set of kinematic transforms. For a system of n robots with q total degrees of freedom, the generalized cluster and robot pose vectors and their kinematic relationships are:

$$\vec{C} \equiv \begin{bmatrix} c_1 \\ c_2 \\ \vdots \\ c_q \end{bmatrix} = \begin{bmatrix} g_1(r_1 \dots r_q) \\ g_2(r_1 \dots r_q) \\ \vdots \\ g_q(r_1 \dots r_q) \end{bmatrix} \tag{1}$$

$$\vec{R} \equiv \begin{bmatrix} r_1 \\ r_2 \\ \vdots \\ r_q \end{bmatrix} = \begin{bmatrix} h_1(c_1 \dots c_q) \\ h_2(c_1 \dots c_q) \\ \vdots \\ h_q(c_1 \dots c_q) \end{bmatrix} \tag{2}$$

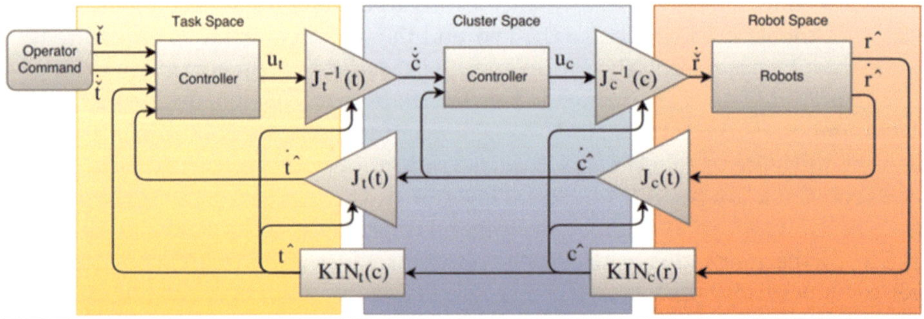

Figure 1.
The multilayer control architecture. Layers exist for task-space, cluster space and robot space control, with inverse Jacobian resolved rate controllers used within each space.

where r_i are robot pose states, c_i are cluster pose states, $g_i(\ldots)$ are kinematic equations, and $h_i(\ldots)$ are inverse kinematic equations [33, 34]. Cluster state velocities are linearly mapped from robot velocities using a Jacobian matrix.

These mathematical transformations are the basis for the layered cluster space control framework shown in **Figure 1**. On the right, in red, are the individual robots, which accept platform-level velocity commands. The blue region is the cluster space controller, shown in the form of a kinematic, resolved rate controller. This layer accepts cluster-level commands regarding cluster mobility and geometry; its outputs are robot velocity commands for each robot in order to achieve the cluster-level goals. Forward kinematic equations are used to compute cluster states from estimated robot states, and cluster control effort is transformed to robot velocity commands using the inverse cluster Jacobian matrix. This control technique has been implemented experimentally in a wide variety of multi-robot systems operating in land, sea, and air [4, 5, 7, 35, 36]. The yellow region of the diagram is described in the next section.

3. Task space control

Just as operational space [37] kinematic transforms are used to allow control of robots based on their actuator configuration, the cluster space methodology uses kinematic transforms to establish the control task in terms of multirobot formation mobility and geometry. Here, we add yet another operational space control later for task-oriented specification of behavior. As shown in **Figure 1**, each control layer uses a resolved rate controller that provides rate commands to an inverse Jacobian function, which converts those commands to rate set-points for the next layer (full dynamic controllers have been demonstrated in other works). Using this approach, operator commands are issued and controlled at the task level, and compensation commands are then successively transformed to cluster velocity, robot velocity and finally actuator velocity set-point commands as the control architecture executes. Each successive layer acts as an inner control loop for the preceding layer.

Given this, the long-range dynamic communication link management system presented in this paper involves two interacting tasks. The first is a "communication space" task that uses a measurement-based, link-balancing control strategy to space robots along the inline dimension between the end stations (a model-based assumption that in-line positioning is best has been made, leading to the use of a simple cross-track nulling controller for the second dimension); furthermore, a measurement-based controller varies the number of robots required by the task to minimally satisfy the desired aggregate communication quality set points. The second task is a simple position control task to maintain unused robots in a ready state. While the second task is trivial, we treat the management and interplay of these multiple distinct tasks in a formal manner.

In total, we are using a multi-layer, multi-task controller. In [24] we develop the conditions to ensure Lyapunov stability for each layer as well as the conditions for task switching. Space limitations prevents their further discussion in this work.

The following subsections demonstrate use of the layered control methodology for the communication and position control tasks. For each task, a formal definition of the layered state spaces is provided, the kinematic and Jacobian transforms used to convert between spaces are established, and controllers are provided for each layer. Both tasks assume the use of planar robots given that this was the type of robot used in the simulations and experiments described in Section VI.

3.1 Example task: long range communication

Consider the task of long-range communications between two exogenous nodes using mobile relays. To maintain the link quality, n_c robotic relay nodes will move to intermediate locations based on desired link characteristics.

3.1.1 Spaces and states

The relevant spaces for this scenario are the individual robot space, the cluster space describing the geometry of the task-specific group of robots, and the communication task space. The robot space is defined by the pose of all robots, specified below for quantity n_c robots:

$$\vec{r} \triangleq \left[x_1, y_1, \theta_1, \ldots, x_{nc}, y_{n_c}, \theta_{n_c}\right]^T \tag{3}$$

where (x_i, y_i) is the Cartesian position and θ_i is the orientation of robot i, assuming planar operation.

The cluster space pose vector describes the location and shape of the cluster. In this case, the separation distances ρ_i and chain angles α_i define the geometry, as depicted in **Figure 2**; although many other cluster definitions are possible, this choice is convenient due to the serial nature of the communications task.

Accordingly, the cluster state vector is defined:

$$\vec{c} \triangleq \left[x_c, y_c, \theta_c, \rho_1, \alpha_1, \phi_1, \ldots, \rho_{n_c-1}, \alpha_{n_c-1}, \phi_{n_c-1}\right]^T \tag{4}$$

In the task space, the user is interested in maintaining sufficient communication quality of service (QoS) between two end nodes, with signals being relayed as needed. Quality of service proved impractical to quantify in real time, so the system measures the link power between nodes using the received signal strength indicator (RSSI). For line-of-sight, the RSSI may be modeled as inversely proportional to the square of the distance between two points, hence:

$$s_i = \frac{k}{(x_{i+1} - x_i)^2 + (y_{i+1} - y_i)^2} = \frac{k}{\rho_i^2} \tag{5}$$

where k is a constant associated with the antenna gain. It is important to note that RSSI is measured directly but this model is used to compute the Jacobian, similar to a model-based approach for gain scheduling.

As depicted in **Figure 3,** the quality of service between the end nodes is influenced by both the crosstrack error, e_{xt}, and the angles of alignment, γ_i. Assuming a line of sight model, the maximum total signal strength is achieved by minimizing the crosstrack error and angles of alignment. The ratio or balance, B_i, of the link power in each segment is also important to avoid data rate bottlenecks or backup in homogeneous systems, or to allow for imbalanced transmission rates in nonhomogeneous

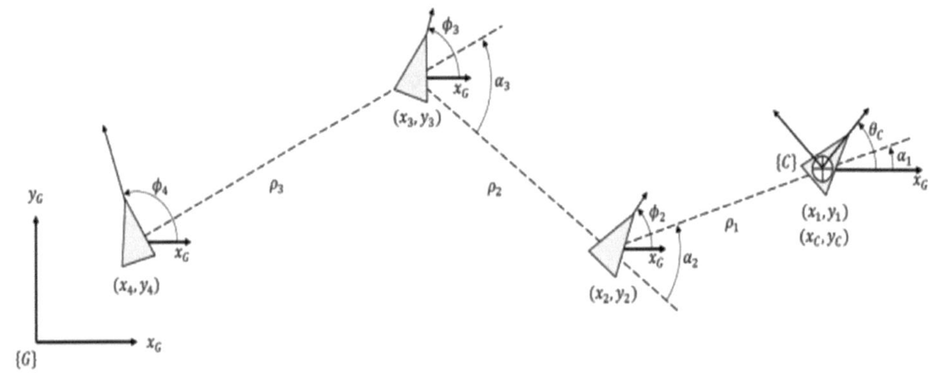

Figure 2.
Serial chain cluster diagram for an i-hop chain.

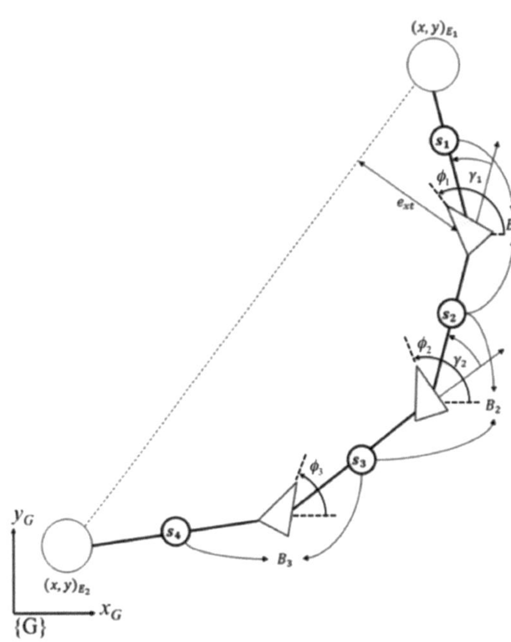

Figure 3.
Long-range link diagram for a 4-hop link performed by 3 robots.

systems. The first and last links are functions of the position of the end nodes being connected $\left(x_{E_1}, y_{E_1}\right), \left(x_{E_2}, y_{E_2}\right)$ which are uncontrolled states of the environment. Lastly, the orientation of the robot, ψ_i, is included to fully define all degrees freedom of system. The communication "pose" vector is defined:

$$\vec{t}\left(x_{E_1}, y_{E_1}, x_{E_2}, y_{E_2}\right) \triangleq \left[B_1, \dots, B_{n_c}, e_{xt}, \gamma_1, \dots, \gamma_{n_c-1}, \psi_1, \dots, \psi_i\right]^T \tag{6}$$

Given these definitions, we define the desired states as follows. For a uniformly balanced network, $B_i = 1$. For minimum crosstrack error for maximum link quality, $e_{xt} = 0$. For aligning the robots between end points, $\gamma_i = 0$. In this case, the robots have a holonomic constraint and so the robot headings, ψ_i, are controlled at the platform level. Trajectory generation is a major topic of research for robotic communication networks [38] but it is not our focus in this work. These commands remain constant throughout each presented experiment, unless otherwise noted. It may be argued that more sophisticated control algorithms require less sophisticated command trajectories.

3.1.2 Kinematic transformation equations

Robot states are transformed into the cluster states using kinematic equations derived from formation geometry:

Cluster frame:

$$x_c \triangleq x_1 \; y_c \triangleq y_1 \; \theta_c \triangleq \theta_1 \tag{7}$$

Chain length:

$$\rho_i \triangleq \sqrt{\left(x_{i+1} - x_i\right)^2 + \left(y_{i+1} - y_i\right)^2} \tag{8}$$

Chain angle:

$$\alpha_i \triangleq atan2\left(y_{i+1} - y_i, x_{i+1} - x_i\right) - \sum_{j=1}^{i-1} \alpha_j \tag{9}$$

Node orientation:

$$\phi_i \triangleq \theta_i \tag{10}$$

where $atan2(\dots, \dots)$. is the two-argument function that calculates a four-quadrant arc tangent with a range of $[\pi, -\pi]$

These cluster states are transformed into the task states using the measured link states and system geometry:

Balance:

$$B_i \triangleq \frac{s_{i+1}}{s_i} = \begin{cases} \dfrac{\left(x_{E1} - x_{c1}\right)^2 + \left(y_{E1} - y_{c1}\right)^2}{\rho_1^2} \text{ for } i = 1 \\ \dfrac{\rho_2^2}{\left(x_{E2} - x_c + \rho_2 \cos(\alpha_1 + \alpha_2) + \rho_1 \cos\alpha_1\right)^2 + \left(y_{E2} - y_c + \rho_2 \sin(\alpha_1 + \alpha_2) + \rho_1 \sin\alpha_1\right)^2} \text{ for } i = n_p - 1 \\ \dfrac{\rho_i^2}{\rho_{i+1}^2} \text{ otherwise} \end{cases} \tag{11}$$

Crosstrack error:

$$e_{xt} = \sqrt{\frac{\left((x_{E_2} - x_{E_1})\left(y_{E_1} - y_c\right) - (x_{E_1} - x_c)\left(y_{E_2} - y_{E_1}\right)\right)^2}{(x_{E_2} - x_{E_1})^2 + \left(y_{E_2} - y_{E_1}\right)^2}} \tag{12}$$

Angle of alignment

$$\gamma_i = \alpha_i \tag{13}$$

Orientation:

$$\psi_i = \phi_i \tag{14}$$

where $\left(x_{E_1}, y_{E_1}\right)$ and $\left(x_{E_2}, y_{E_2}\right)$ are the positions of the end stations that are being connected by the multi-robot communication system.

3.1.3 Jacobian matrices

The Jacobian matrices are computed from the kinematic equations to map velocities between spaces. The solution is typically lengthy and so not shown here but easily computed using (5) with (7–10) for the cluster Jacobian or with (11)–(14) for the task Jacobian.

3.1.4 Control design and performance

In addition to determining the kinematic transforms, control laws must also be formulated. For these experiments, simple linear controllers suffice as the testbed platforms have well-behaved dynamics, but the system architecture can accommodate any type of control algorithm within each space.

As many commercially available robotic platforms control their own local velocity, a detailed discussion of platform control is not necessary. It is important to note that if the robots are nonholonomic, orientation and global translation are coupled which precludes independent control of all degrees of freedom. For details on our testbed nonholonomic control, please see [36].

The following equations can be used to design controllers in higher spaces using traditional techniques. The transfer functions at each layer can be approximated as linear, time-invariant (LTI) with proper tuning, maintaining diagonal dominance, and avoiding singularities. General system stability and performance is discussed in Section III.C.

Cluster space response:

$$\dot{\vec{c}} = \left(J_c^{-1} + G_r J_c^{-1} H_c\right)^{-1} G_r J_c^{-1} H_c \dot{\vec{c}}_d = G_c \dot{\vec{c}}_d \tag{15}$$

Task space response:

$$\vec{t} = \left(J_t^{-1} sI + G_c J_t^{-1} H_t\right)^{-1} G_c J_t^{-1} H_t \check{t} = G_t \vec{t}_d \tag{16}$$

where G_x represents a diagonal matrix of transfer functions in space x, H_x represents a controller in space x. The system pose is represented by r in robot space, c in cluster space, and t in task space. As subscripts, these letters associate the variable with a space. Subscript d denotes desired states.

For these particular experiments, the cluster space control law utilizes proportional feedforward and feedback, shown below, for response time and error rejection respectively:

$$\vec{u}_c = H_c\left(\dot{\vec{c}}_d, \vec{c}\right) = K_{c_f}\dot{\vec{c}}_d + K_{c_p}\left(\vec{c}_d - \vec{c}\right) \tag{17}$$

where u_c denotes cluster space control effort, $\dot{\vec{c}}_d$ denotes desired cluster velocity, K_{c_f} denotes proportional feedforward gain matrix, and K_{c_p} denotes proportional feedback gain matrix.

For these particular experiments, the communication task space uses proportional feedback control, shown below:

$$\vec{u}_t = H_t\left(\vec{t}_d, \vec{t}\right) = K_{t_p}\left(\vec{t}_d - \vec{t}\right) \tag{18}$$

where K_{t_p} is the feedback gain matrix and $\vec{t}_d$ is the desired state. These desired states are discussed in Section III.A.1). While simplistic, these control laws yield sufficient performance in this application and our prior work with different tasks performed in land, sea, and aerial environments [4, 5, 7, 12, 39].

3.2 Example task: position control

In this task, n_p robots are tasked to go to and maintain a specified position. Since the task is a direct specification of individual robot positions, the task is identical to many one-robot clusters which is identical to control of individual robots. While this degenerates into trivial task, we adhere to the layered control architecture in order to provide unified control of all mission-related robots.

Space limitations prevent a complete description of this task. Given it's simplicity, key aspects of its implementation are reviewed here. To begin, the robot cluster and task pose vectors are:

$$\vec{r} \triangleq \left[x_1, y_1, \theta_1, \dots, x_{n_p}, y_{n_p}, \theta_{n_p}\right]^T \tag{19}$$

$$\vec{c} \triangleq \left[x_{c_1}, y_{c_1}, \theta_{c_1}, \dots, x_{c_{np}}, y_{c_{np}}, \theta_{c_{np}}\right]^T \tag{20}$$

$$\vec{t} \triangleq \left[x_{t_1}, y_{t_1}, \theta_{t_1}, \dots, x_{t_{np}}, y_{t_{np}}, \theta_{t_{np}}\right]^T \tag{21}$$

where (x_i, y_i, θ_i) is the robot i pose, $\left(x_{c_i}, y_{c_i}, \theta_{c_i}\right)$ is the cluster pose for cluster i, and $\left(x_{t_i}, y_{t_i}, \theta_{t_i}\right)$ is the task-level pose for robot i. For the position control task, the cluster positions and the task positions are both equated to the robot positions. Accordingly, the forward and inverse kinematic relationships are unity, and the Jacobians are the unit matrix.

For our experiments, the cluster space velocity control law utilizes proportional feedforward and feedback, shown below, for response time and error rejection respectively:

$$\vec{u}_c = K_{c_f}\dot{\vec{c}}_d + K_{c_p}\left(\dot{\vec{c}}_d - \dot{\vec{c}}\right) \tag{22}$$

where $\vec{u}_c$ denotes cluster space control effort, $\dot{\vec{c}}_d$ denotes the desired cluster velocity, K_{c_f} denotes a proportional feedforward gain matrix, and K_{c_p} denotes a proportional feedback gain matrix. Similarly, the task-space state controller utilizes proportional feedback for error rejection:

$$\vec{u}_t = K_{t_p}\left(\vec{t}_d - \vec{t}\right) \tag{23}$$

where $\vec{u}_t$ denotes task space control effort, $\vec{t}_d$ denotes the desired task state and K_{t_p} denotes a proportional feedback control gain.

4. Multi-tasking missions

In the previous section, an approach for developing a layered task control architecture for the motion requirements of a multi-robot task is presented. Our interest, however, is not only in task-oriented control of a robot group but also in conducting activities that require multiple interacting tasks, each potentially implemented by a multi-robot group. We term the conduct of such an activity a "mission." Furthermore, we are interested in collaborative multi-task missions, with "collaboration" implying the ability for one task to support another and for tasks to share resources, such as robots, in an appropriate manner.

To achieve this vision, **Figure 4** shows a control architecture that integrates the two tasks required for the communication relay mission. The architecture integrates the operation of the tasks in two ways. First, on the "front end" of the architecture, a mission-level specification interface provides a mechanism for defining mission-oriented objectives and assigning them to the tasks. In addition, the tasks are allowed to interact, providing a mechanism for tasks to issue commands to and to set constraints for each other. For the communication relay mission, the communication task acts as a master task in order to determine the number of robots it requires; once this determination is made, it specifies the number of remaining robots (to assign to a position hold task) to the position task for position control. In ongoing work with more complex missions, the robot allocation process is independent of any one task. Second, on the "back end" of the architecture, unified motion control is enabled by consolidating the control elements of multiple tasks as described in the following subsection.

Multi-agent systems have the capacity for functional diversity which motivates the ability to change roles during operation. As such, an additional focus of this work investigates the process of reallocating autonomous agents within the given framework. Doing so requires (A) a framework for supporting and transitioning multiple clusters within the same architecture, (B) defining allocation policies to determine when and how to reallocate resources between tasks, and (C) analytic methods to safely transition between robot configurations.

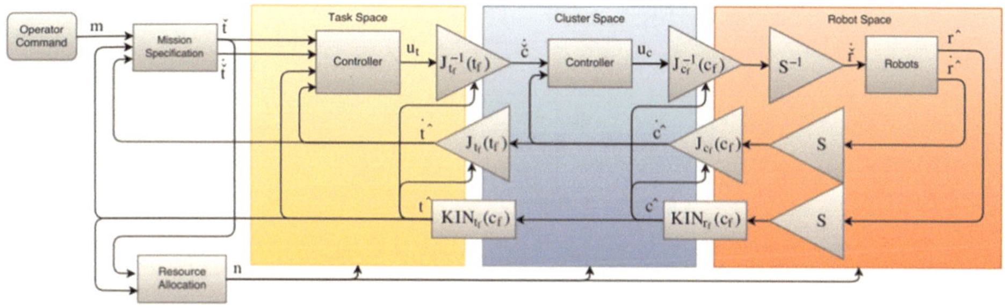

Figure 4.
A multi-task collaborative Mission control architecture. The diagram shows the consolidated control architecture with additional functions to provide mission specification and resource allocation.

4.1 Representation framework

Simultaneously accomplishing multiple mobility tasks requires assigning different tasks to different robots, which, in our framework, implies the use of multiple task-specific robot clusters. While multiple instances of the control framework of Section III could be run in parallel, this approach is static and unable to conveniently support the reconfiguration of clusters as robots are reallocated. Instead, the single-task multi-layered framework is extended to become a multi-task controller that operates on a "federated" system state vector. At a given moment in time, given an assignment of a specific number of robots to specific tasks as defined by the robot allocation vector in (39), the federated kinematic equations and federated Jacobian matrices are composed of the cluster-specific kinematic equations and Jacobian matrices, as shown in (41) and (42).

Robot allocation vector:

$$\vec{n} = [n_1, n_2, \ldots, n_o]^T \text{where } n = \sum_{i=1}^{o} n_i \tag{24}$$

Federated pose vectors:

$$\vec{r}_f \triangleq \left[\vec{r}_1, \vec{r}_2 \ldots, \vec{r}_o\right]^T \tag{25}$$

$$\vec{c}_f \triangleq \left[\vec{c}_1, \vec{c}_2 \ldots, \vec{c}_o\right]^T \tag{26}$$

$$\vec{t}_f \triangleq \left[\vec{t}_1, \vec{t}_2 \ldots, \vec{t}_o\right]^T \tag{27}$$

Federated kinematic equations:

$$KIN_{c_f}\left(\vec{c}_f\right) \triangleq \left[KIN_{c_1}\left(\vec{r}_1\right), KIN_{c_2}\left(\vec{r}_2\right), \ldots, KIN_{c_o}\left(\vec{r}_o\right)\right]^T \tag{28}$$

$$KIN_{t_f}\left(\vec{t}_f\right) \triangleq \left[KIN_{t_1}\left(\vec{c}_1\right), KIN_{t_2}\left(\vec{c}_2\right), \ldots, KIN_{t_o}\left(\vec{c}_o\right)\right]^T \tag{29}$$

Federated Jacobian matrices:

$$J_{c_f} \triangleq \begin{bmatrix} J_{c_1} & \cdots & 0 \\ \vdots & \ddots & \vdots \\ 0 & \cdots & J_{c_o} \end{bmatrix} \tag{30}$$

$$J_{c_f} \triangleq \begin{bmatrix} J_{c_1} & \cdots & 0 \\ \vdots & \ddots & \vdots \\ 0 & \cdots & J_{c_o} \end{bmatrix} \tag{31}$$

where subscript f denotes federated elements; n_i is the number of robots assigned to task i; $\vec{r}_i, \vec{c}_i$ and $\vec{t}_i$ are the robot space, cluster space, and task space pose vectors for task i; $KIN_{c_i}(\vec{r}_i)$ and $KIN_{t_i}(\vec{c}_i)$ are the cluster space and task space kinematic equations for task i; J_{c_i} and J_{t_i} are the cluster space and task space Jacobian matrices for task i; and o is the number of tasks spanning the multi-robot system. The federated pose vector is formulated by concatenating pose vectors, and similarly for the federated kinematic equations. The federated Jacobian matrix is block-diagonal, comprised of uncoupled Jacobian matrices. As the agents shift between tasks, individual elements change size, but the size of the federated elements and the overall structure of the mission-level control system remain constant.

As a simple example of multitasking, consider combining the two previously described tasks into the following two-task mission: maintain communication between two end points or otherwise move to an idle parking position. A subset of the federated elements are shown below for two configurations of a $n = 3$ robot system: (1) one robot is allocated to the communications task and two robots are idle ($\vec{n} = [n_{com}, n_{idle}]^T = [1,2]^T$) and (2) two robots are allocated to the communications task and one robot is idle ($\vec{n} = [n_{com}, n_{idle}]^T = [2,1]^T$):

Federated cluster pose vector:

$$\vec{c}_M = \begin{cases} \left[[x_c, y_c, \theta_c] \left[x_{I_1}, y_{I_1}, \theta_{I_1}, x_{I_2}, y_{I_2}, \theta_{I_2} \right] \right]^T \text{for } \vec{n} = [1,2]^T \\ \left[[x_c, y_c, \theta_c, \rho_1, \alpha_1, \phi_1] \left[x_{I_1}, y_{I_1}, \theta_{I_1} \right] \right]^T \text{for } \vec{n} = [2,1]^T \end{cases} \tag{32}$$

4.2 Allocation policies

Given the general desire to accommodate multiple tasks and the ability to reassign robots between tasks, we need to incorporate an allocation function for assigning specific robotics to specific tasks. The allocation problem is well studied [38, 40, 41], and we are not proposing any particular innovations in this area; rather, our interest is in determining how any such allocation policy fits into the proposed architecture. To date, we believe that the necessary interface consists of providing the allocation policy with the commanded and actual task state vector such that it can compute $\vec{n}$, which is consistent with the approaches discussed in [38]. This value is then provided to each layer of the control architecture in order for controllers and kinematic transforms with appropriate internal dimensions to be selected.

To explore this set of interfaces, we have adopted a simple-but-common state-machine based allocation policy for the communication-idle task allocation process. In particular, we define an aggregate QoS metric known as the chain capacity [27]:

$$\delta \triangleq \frac{n_{com} + 1}{\sum_{i=1}^{n_{com}+1} \frac{1}{s_i}} \tag{33}$$

which we use for the transition function:

$$[n_1 \ n_2] = \begin{cases} [n_1 + 1 \quad n_2 - 1], \text{if } \delta < k_1 \\ [n_1 - 1 \quad n_2 + 1], \text{if } \delta > k_2 \\ [n_1 \quad n_2], \text{otherwise} \end{cases} \tag{34}$$

where k_1 and k_2 are performance thresholds that together provide switching hysteresis. For the communication systems and scenarios explored later in this paper, k_1 and k_2 were set to values equivalent to the line-of-sight link power at 40 m and 10 m, respectively. For example, if $n = [2 \quad 1]$ and the signal strength was measured $s = [-75 \quad -76 \quad -78]^T$dBm giving $\delta = -76.4$dBm, and the thresholds were $k = [-52 \quad -76]^T$dBm, then the next robot allocation would be $n = [3 \quad 0]$.

To reiterate, this policy sets the number of robots to achieve a given link quality while the task-level control provides link balance as stated in (11). This policy is appropriate for the communication relay application and it is also consistent with the general interface requirements hypothesized in the previous paragraph. Future researchers can incorporate the allocation policy that is most appropriate for their tasks, whether state of the art techniques (see [38]) or commonplace yet effective techniques like state machines.

5. Experimental testbed

Experimental work used the proven [35] SCU multi-robot test bed, with a communication relay payload. This student-developed system consists of several Pioneer 3-AT skid steered robots with a custom suite of avionics and a centralized off-board control workstation. Wireless 28.8 kbps Ricochet modems are used to relay robot drive commands and position data between the control workstation and the robots. BasicX microcontrollers route drive commands to the robot's built-in speed controller and collect data from a Garmin GPS18 unit and a Devantech CMPS10 compass. Based on experimental evaluation, robot velocity dynamics are approximated as a second-order system with $\zeta = 0.7$ and $\omega_n = 2\pi 0.25$ rad/s; simulations used these speed response characteristics for each robot.

Within the control workstation, an open source real-time data streaming server, known as the DataTurbine, relays information between MATLAB/Simulink and simple applications that handle serial port data flow to/from the wireless modems. Controllers execute in real-time within Simulink; this promotes rapid, iterative development in the field and supports rudimentary operator interfaces. Using a dual-core laptop computer, the system maintains a 5 Hz servo loop rate.

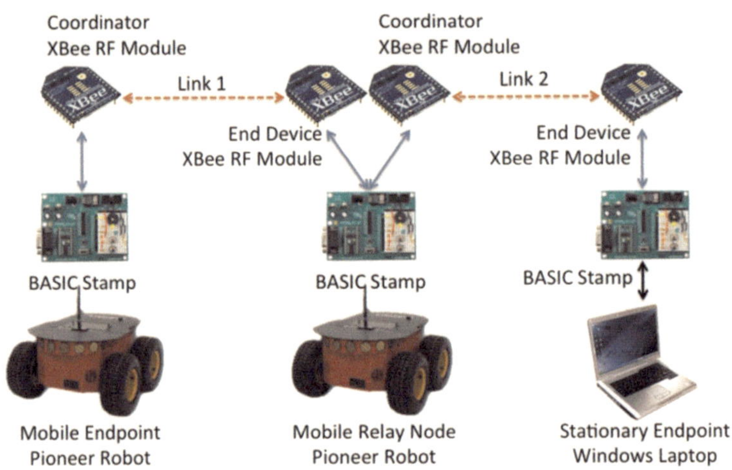

Figure 5.
Components in the multi-robot communications relay test bed.

Each robot carries a communications relay payload comprised of two Digi International XBee Series 2 wireless transceivers connected by a BASIC Stamp microcontroller, shown in **Figure 5**. The microcontroller appends the received signal strength indicator (RSSI) value to each message prior to retransmission. When the end station ultimately receives the message, it also obtains the RSSI state for the multi-hop link. This state data is provided to the task layer controller in order to determine how to adjust the position of the relay cluster. Messages and the associated RSSI measurements were generally executed at 1 Hz, but dropouts due to the inexpensive hardware often resulted in short periods of slower execution, adding a realistic challenge to the experiment.

6. Experimental results

A number of simulations and experiments were executed to demonstrate functionality of the system and to showcase particular advantages of the control architecture. First, two scenarios were examined with the single communications task: (A) an experiment showing the single robot system response to hardware configuration changes such as reductions in transmission power, and (B) a simulation showing multi-robot system response to a mobile end station and environmental attenuation. Next, three scenarios were examined performing multiple tasks of communications and idle position control, thereby allowing robots to be added or removed from the communication task: (C) an experiment showing responses to desired link quality commands that require robot reallocation, (D) a simulation showing system response to a moving mobile end station in which robot reallocation is required, and (E) an experiment showing the same capability as in D. Experimental work has demonstrated that the architecture is tolerant of real-world phenomena such as sensor noise, quantization, model mismatches, and communication delays. Simulations allowed rapid exploration with higher numbers of robots, allowing clear demonstration of the architecture behavior without hardware constraints.

6.1 Experiment: single task single robot behavior with hardware configuration change

This scenario examines the control system response to configuration changes such as component degradation or a power reduction used to conserve energy. One robot is used to relay communications between two fixed end stations. The system is commanded to achieve unity link balance with null crosstrack error. In this experiment, the system initially moves to an equilibrium position given a nominal communications configuration. Then, at t = 800 sec, the transmission power of end station E_2 is reduced, and the relay robot moves to achieve link equilibrium.

An overhead view of robot position is shown in **Figure 6** where each subplot corresponds to a different time window; in the first, the robot moves to an equilibrium position, and in the second, the robot adjusts its position to balance the link given the power reduction at station E_2. In **Figure 7** the RSSI values and the link balance parameter (commanded to 1) are shown. In each phase, the systems moves to achieve the commanded link balance, resulting in balanced RSSI values.

6.2 Simulation: single task multi-robot behavior with a mobile end station and local attenuation

This scenario evaluates system behavior given local attenuation effects such as obstructions, fog, or foliage. Three robots are used to relay communications between a fixed base station and a mobile end-node. A comparison is made of the system's performance with and without measurement compensation for these effects to demonstrate how our communications task improves performance compared to the use of a simple link model.

A overhead view is shown in **Figure 8** with robot trajectories plotted for both ideal and attenuated scenarios. Messages are relayed between the base node, located at the origin of the plot, and the mobile end-node, which has a quarter-circle trajectory plotted in black and running from (60,0) to (0,60). A region of power attenuation exists for $y > 40$, where any link involving a robot within this area is reduced by half. As the remote end-node traverses its arc at a constant speed, a three-robot cluster maintains link balance as described before. In the ideal case, the robots spread evenly and follow the traverse in concentric arcs, consistent with a model-based approach. In

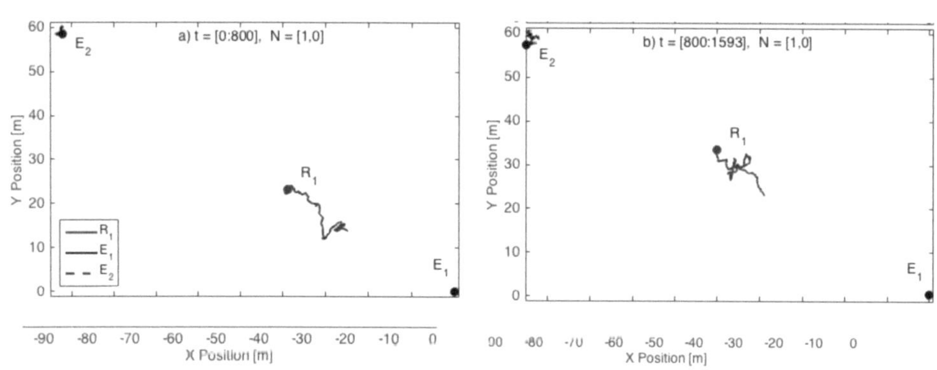

Figure 6.
Overhead view of positions of robot $\mathbf{R_1}$ and fixed end nodes $\mathbf{E_1}$ and $\mathbf{E_2}$ at specified times during hardware configuration change experiment.

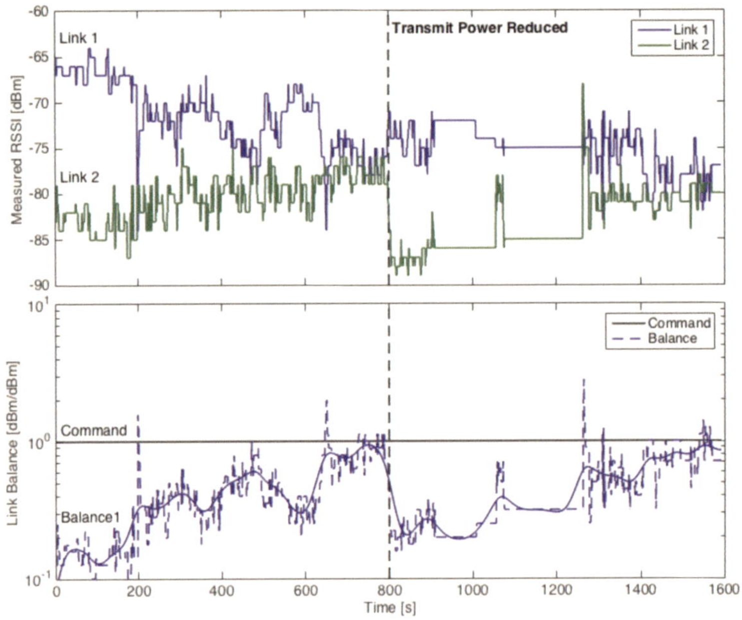

Figure 7.
Time history of link power and balance ratio during hardware configuration change experiment.

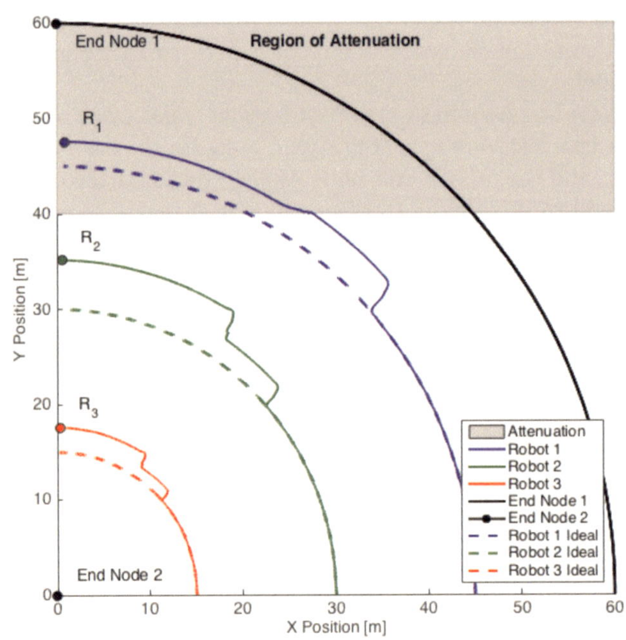

Figure 8.
Overhead view of robot ***R*** *positions comparing trajectories in ideal transmission environments (dashed) and trajectories responding to an encountered region of attenuation (solid).*

the non-ideal case with attenuation, the multi robot system begins as before, but alters its trajectory to rebalance the links when it senses a drop in signal strength as nodes enters the region of attenuation.

DOI: http://dx.doi.org/10.5772/intechopen.111825

This example demonstrates the value of direct measurement of communication states and high-level task-space control. Sensing the signal strength allows the system to maintain the desired state despite unanticipated characteristics of the environment. In contrast, an open-loop, model-based approach would evenly distribute the nodes as shown, yielding lower performance in non-ideal environments.

6.3 Experiment: multi-task multi-robot behavior with configuration link quality command response

This scenario demonstrates changing user requirements for better connectivity or higher throughput thereby forcing a change in the cluster configuration. With fixed communication endpoints and an increase in the commanded link quality, two robots are sequentially reallocated from the idle task to the communication relay task. Each newly incorporated robot moves from its idle position to a location determined through execution of the communication task in order to achieve the commanded link quality and link balance set-points while the other robots in the communication task reposition themselves accordingly.

Figure 9 shows the fixed end nodes E_1 and E_2 and two mobile robots. In the top plot, for time $t = [0:155]$, the commanded link quality is such that the two end

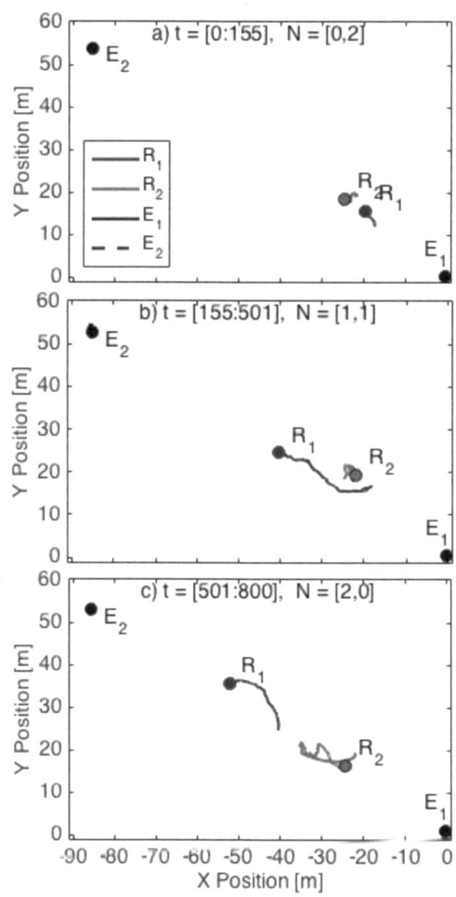

Figure 9.
Overhead view of positions of robots **R** *in response to a link quality command between end nodes* **E**.

stations communicate directly with each either without the need for a relay node. As a result, the configuration is $\vec{n} = [0, 2]^T$, with both robots assigned to an idle task and holding their position. As seen in **Figure 10**, link quality is maintained within an acceptable deadband; the link balance is not shown given the single hop.

For time $t = [155 : 501]$, the link quality setpoint is increased. As seen in **Figure 10b**, the link quality deadband is violated, leading to a change in the assignment of tasks such that $\vec{n} = [1, 1]^T$, shown in **Figure 10a**. This results in controlling B_1, the link balance between the two existing sub-links, by achieve balance as shown in **Figure 10c** by the repositioning of robot 1, as shown in **Figure 9b**.

During time $t = [501 : 800]$, the system responds to another command to increase link quality. Again, the link quality deadband is violated, leading the addition of robot R_2 to the communication task. With two intermediate robots, two link balancing operations take place in parallel, shown in **Figure 10c**. The initial position of the added robot creates a large switching transient in the link balance. Accordingly, the robots alter their positions, as shown in **Figure 9c**.

The plots in **Figure 10** show that the sensed RSSI parameters were clearly not ideal, exhibiting noise and quantization and the effects of a wide range on non-ideal characteristics of the wireless links. Because of these non-ideal characteristics, the robots do not move to the geometric center of the end points. These real-world phenomena are challenging but the control architecture is sufficiently robust to tolerate these unmodeled effects.

6.4 Simulation: multi-task multi-robot behavior with a mobile end station

This simulation demonstrates control of link quality and balance with a mobile endpoint, gracefully adding and removing robots as appropriate for the task. The

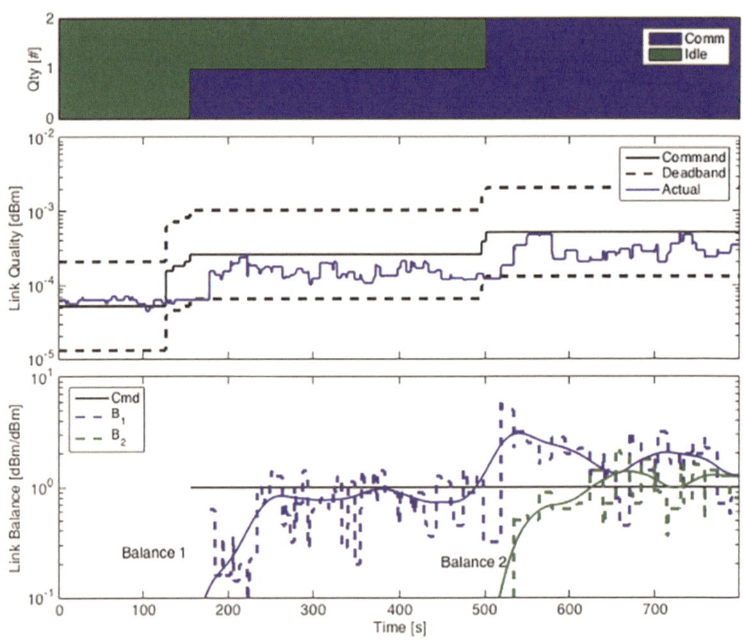

Figure 10.
Time history of key states while evaluating the link quality commanded response, forcing configuration change.

scenario starts with a mobile end station communicating directly with a fixed end station and with five robots executing the idle position hold task. As the mobile end station executes an elliptical trajectory that moves it away from and then closer to the fixed end station, the five relay robots are sequentially added and then removed to the communication task, maintaining the specified link quality and balance (**Figure 11**).

The evolution of the system's state can be seen in **Figure 12**. For time t = [0,500], the end station moves away from the fixed end station, lowering link quality, as shown in **Figure 12b**. At times t = ~110 sec, 193 sec, 265 sec, ~340 sec and ~ 490 sec, the link quality hits the threshold for acceptable link quality. At each of these times, the controller re-assigns a new robot from the idle task to the communications task, thereby causing $\vec{n}$ to change at these times, as seen if **Figure 12a**, and new robots being added to the multi-hop link, as seen in **Figure 12c**; idle robots are controlled to remain in their default position. These additions to sub-links in the communication chain lead to new balance parameters to be controlled, B_1 through B_5.

As the mobile end station turns back towards the base node, the link quality increases. Each time this value hits the high deadband, at times t = 845 sec, ~870 sec, ~890 sec, ~910 sec, and ~ 935 sec, an active communication task robot is returned to the idle task. This reduces the number of link balance parameters, leading to a transient in link balance that is quickly controlled through the repositioning of the remaining communication task robots. Interestingly, the deadband causes unequal times between transitions as the robots are faster to move out due to the task state definition and allocation policy and slower to move into the communication cluster.

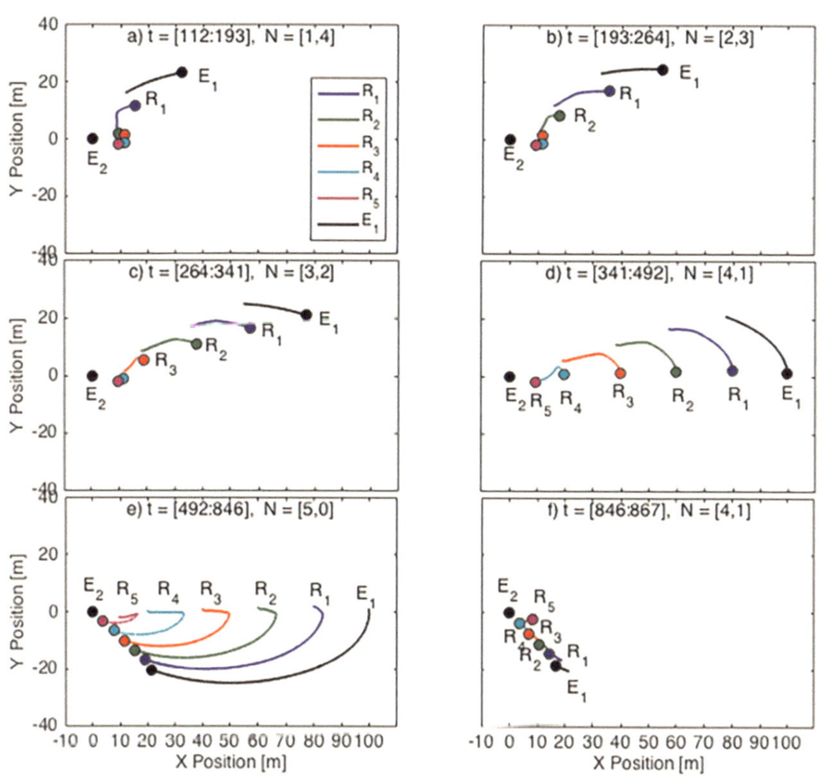

Figure 11.
*Overhead view of robots **R** and end nodes **E** during specified times for mobile endpoint simulation.*

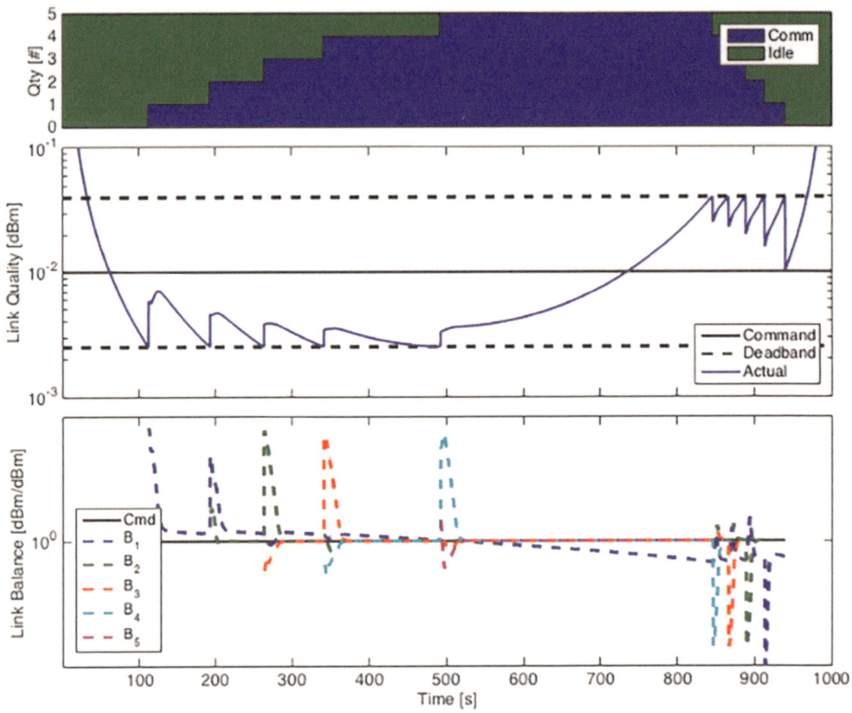

Figure 12.
Time history of key system states for mobile endpoint simulation.

This demonstrates the ability of the control architecture to respond to motion of the end node based on sensed link characteristics and to reallocate robots without any additional commanding.

6.5 Experiment: Multi-task multi-robot behavior with a mobile end station

Like the simulation presented in Section VI.D, this experiment demonstrates the control of link quality and balance with a mobile end node. The experiment starts with the end stations near each other and directly communicating, with two relay robots in an idle position. As the mobile end station moves away, relay robots are sequentially added to maintain the specified level of link quality and balance.

Figure 13 shows the paths taken by the robots and endpoints, and **Figure 14** shows the corresponding state trajectories. In **Figure 13a**, for time $t = [0 : 148]$, the mobile endpoint can be seen moving away from moving away from the stationary endpoint while the link quality remains within the deadband. The communication relay robots are allocated to idle, $\vec{n} = [0, 2]^T$, and can be seen parking themselves.

At time $t = 148$, the link quality exceeds the lower bounds of the deadband and the allocation policy adds a robot to the communication relay task, changing the configuration vector to $\vec{n} = [1, 1]^T$. In **Figure 13b**, for time $t = [148 : 591]$, the new robot relay moves to balance the communication links while the mobile end station continues moving away from the stationary endpoint. Though there is not significant movement of the relay robot, the measured link states, shown in **Figure 14**, indicate that the balance setpoint is achieved during this time. This demonstrates the complexity and non-intuitiveness of RF fields and the benefit of communication-space measurement

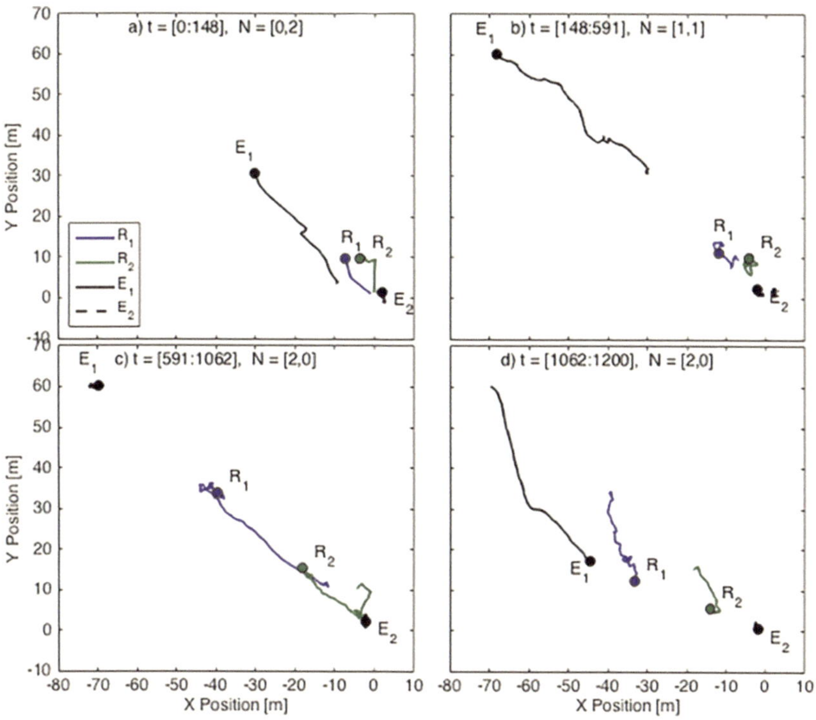

Figure 13.
Overhead view of robots **R** *and end nodes* **E** *during specified times for mobile endpoint experiment.*

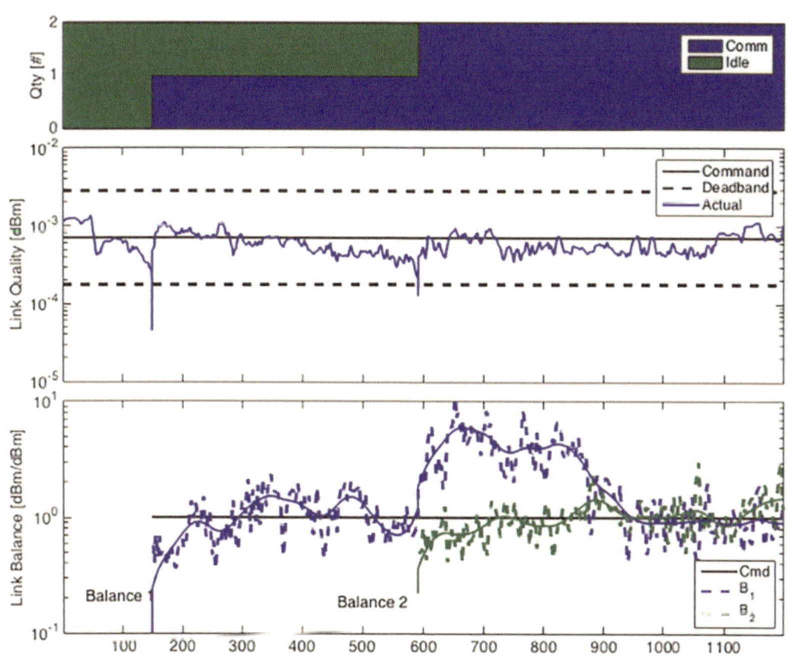

Figure 14.
Time history of key system states for mobile endpoint experiment.

and control, including the effects of the deadbands; alternatively locating the relay robots in the geometric center of the two points would yield worse performance.

At time $t = 591$, the link quality again exceeds the lower bounds of the deadband and the allocation policy adds the second robot to the communication relay task, changing the configuration vector to $\vec{n} = [2, 0]^T$. In **Figure 13c**, for time $t = [591 : 1062]$, both robots move to balance the communication links. The switching transient can be seen in **Figure 14** starting at t = ~600 sec and settling by t = ~950 sec. The final overhead plot, **Figure 13d**, shows the mobile endpoint arcing back towards the stationary endpoint and the relay robots mimic its motion to maintain link balance.

7. Future work and summary

In this article, we presented a multi-robot control architecture providing explicit task control for improved performance yet with abstraction from implementation. Direct sensing and operational task space control eliminate errors due to modeling and implicit specification. In addition, controllers can compensate for non-ideal behavior in the appropriate space of the layered architecture. Abstraction provides the flexibility to engage different types and quantities of robots to accomplish tasks. This segregates individual task complexity in order to facilitate large-scale missions with many tasks. We proposed a design methodology for composing new spatially-sensitive tasks that includes conditions for stability and quantification of responsiveness. The architecture was demonstrated using the example task of communication. Experiments and simulations exhibited explicit control of task states, compensating for the complex behavior inherent in real-world communication networks. The system successfully reacts to a dynamic environment, varying operator commands, and hardware configuration changes.

Ongoing work leverages this architecture for larger missions comprised of more tasks with more complex interactions. This article is a step towards multi-task missions indicative of systems of systems. Continued development of this architecture with new applications and new environments increases the utility of integrated multi-agent systems.

Author details

John Shepard and Christopher Kitts*
Santa Clara University, Santa Clara, CA, USA

*Address all correspondence to: ckitts@scu.edu

References

[1] Joordens MA, Jamshidi M. Consensus control for a system of underwater swarm robots. IEEE Systems Journal. 2022;**4**(1):65-73

[2] Balch T, Arkin RC. Behavior-based formation control for multi-robot teams. IEEE Transactions on Robotics and Automation. 1998;**14**(6):926-939

[3] Ailon A, Zohar I. Control strategies for driving a group of nonholonomic kinematic mobile robots in formation along a time-parameterized path. IEEE/ASME Transactions on Mechatronics. 2012;**17**(2):326-336

[4] Mahacek P, Kitts C, Mas I. Dynamic guarding of marine assets through cluster control of automated surface vessel fleets. IEEE/ASME Transactions on Mechatronics. 2012;**17**(1):65-75

[5] Mas I, Li S, Acain J, Kitts CA. Entrapment/escorting and patrolling missions in multi-robot cluster space control. In: IEEE/RSJ International Conference on Intelligent Robots and Systems; St. Louis, USA. New York City, NY, USA: IEEE; 2009

[6] Elston J, Frew EW. Hierarchical distributed control for search and tracking by heterogeneous aerial robot networks. In: IEEE International Conference on Robotics and Automation; Pasadena, CA, USA. New York City, NY, USA: IEEE; 2008

[7] Adamek T, Kitts C, Mas I. Gradient-based cluster space navigation for autonomous surface vessels. IEEE/ASME Transactions on Mechatronics. 2014;**20**(2):506-518

[8] Li S, Kong R, Guo Y. Cooperative distributed source seeking by multiple robots: Algorithms and experiments. IEEE/ASME Transactions on Mechatronics. 2014;**19**(6):1810-1820

[9] Jevtic A, Gutierrez A, Andina D, Jamshidi M. Distributed bees algorithm for task allocation in swarm of robots. IEEE Systems Journal. 2012;**6**(2):296-304

[10] Ross A, Layton T, Romanishin J, Rus D. IkeaBot: An autonomous multi-robot coordinated furniture assembly system. In: IEEE International Conference on Robotics and Automation; Karlsruhe, Germany. New York City, NY, USA: IEEE; 2013

[11] Yoon D-K, Seo J-T, Yi B-J. Automatic lighting system using multiple robotic lamps. IEEE/ASME Transactions on Mechatronics. 2014;**19**(3):963-974

[12] Okamoto G, Chen C, Kitts C. Beamforming performance for a reconfigurable sparse array smart antenna system via multiple mobile robotic systems. In: Proc of the SPIE - The International Society for Optical Engineers. 2010

[13] Tortora G, Dario P, Menciassi A. Array of robots augmenting the kinematics of endocavitary surgery. IEEE/ASME Transactions on Mechatronics. 2014;**19**(3):1821

[14] Zemiti N, Morel G, Ortmaier T, Bonnet N. Mechatronic design of a new robot for force control in minimally invasive surgery. IEEE/ASME Transactions on Mechatronics. 2007;**12**(2):143-153

[15] Ansola PG, Higuera AG, Otamendi FJ, Morenas J. Agent-based distributed control for improving complex resource scheduling: Application to airport

ground handling operations. IEEE Systems Journal. 2014;**8**(4):1145-1157

[16] Grocholsky B, Keller J, Kumar V, Pappas G. Cooperative air and ground surveillance. IEEE Robotics and Automation Magazine. 2006;**2006**:16-26

[17] Kudelski M, Gambardella LM, Di Caro GA. A mobility-controlled link quality learning protocol for multi-robot coordination tasks. In: 2014 IEEE International Conference on Robotics and Automation (ICRA). 2014. pp. 5024-5031

[18] Kim J, Ladosz P, Oh H. Optimal communication relay positioning in mobile multi-node networks. Robotics and Autonomous Systems. 2020;**129**: 103517

[19] Pezeshkian N, Nguyen HG, Burmeister A. Unmanned ground vehicle radio relay deployment system for nonline-of-sight operations. In: Proceedings of the 13th IASTED International Conference on Robotics and Applications (RA '07). San Diego CA, USA: Space and Naval Warfare Systems Center; 2007. pp. 501-506

[20] Ollero A, Marron PJ, Bernard M, et al. AWARE: Platform for autonomous self-deploying and operation of wireless sensor-actuator networks cooperating with unmanned AeRial vehiclEs. In: Proc of IEEE International Workshop on Safety, Security and Rescue Robotics (SSRR '07). New York City, NY, USA: IEEE; 2007. pp. 1-6

[21] Pezeshkian N, Neff J, Hart A. Link quality estimator for a mobile robot. In: Int. Conf. on Informatics in Control, Automation and Robotics; Rome, Italy. 2012

[22] Mehta VK, Arrichiello F. Ithica, NY, USA: Cornell University; 2013. [Online]. http://arxiv.org/abs/1312.2526v1

[23] Sweeney J, Grupen R, Shenoy P. Active QoS flow maintenance in controlled mobile networks. In: The Fourth International Symposium on Robotics and Automation (ISRA); Queretaro, Mexico. 2005

[24] Shepard J, Kitts C. A multi-robot control architecture for collaborative missions comprised of tightly- coupled, interconnected tasks. IEEE Systems Journal. 2016;**12**(2):1435-1446

[25] Gao Y, Chen H, Li Y, Lyu C, Liu Y. Autonomous Wi-Fi relay placement with mobile robots. IEEE/ASME Transactions on Mechatronics. 2017;**22**(6):2532-2542

[26] Im H-J, Lee C-E, Cho Y-J, Sunghoon K. RSSI-based control of mobile cooperative robots for seamless networking. In: Control, Automation, and Systems, International Conference on; Jeju Island, Korea. New York City, NY, USA: IEEE; 2012

[27] Dixon C, Frew EW. Optimizing cascaded chains of unmanned aircraft acting as communication relays. IEEE Journal on Selected Areas in Communications. New York City, NY, USA: IEEE; 2012;**30**(5):883-898

[28] Bryan J. Thibodeau, Andrew H. Fagg, and Brian N. Levine, "Signal Strength Coordination for Cooperative Mapping; Amherst, USA, 2005.

[29] Min BC, Kim Y, Lee S, Jung J, Matson ET. Finding the optimal location and allocation of relay robots for building a rapid end-to-end wireless communication. Ad Hoc Networks. Amsterdam, Netherlands: Elsevier; 2016; **39**:23-44

[30] Stephan J, Fink J, Kumar V, Ribeiro A. Concurrent control of mobility and communication in multirobot systems.

IEEE Transactions on Robotics. 2017;**33**(5):1248-1254

[31] Bezzo N, Yan Y, Fierro R, Mostofi Y. A decentralized connectivity strategy for mobile robot swarms. In: 18th World Congress of the International Federation of Automatic Control; Milan, Italy. New York City, NY, USA: IFAC; 2011. pp. 4501-4506

[32] Bezzo N et al. A cooperative heterogeneous mobile wireless mechatronic system. IEEE/ASME Transactions on Mechatronics. New York City, NY, USA: IEEE; 2014;**19**(1): 20-31

[33] Kitts C, Mas I. Cluster space specification and control of mobile multirobot systems. IEEE/ASME Transactions on Mechatronics. 2009;**14**(2):207-218

[34] John J. Craig, Introduction to Robotics: Mechanics and Control. 3rd ed. Pearson Prentice Hall; 2005

[35] Mas I, Petrovic O, Kitts C. Cluster space specification and control of a 3-robot mobile system. In: Robotics and Automation, 2008. ICRA 2008. IEEE International Conference on; Pasadena, CA, USA. 2008. pp. 3763-3768

[36] Mas I, Kitts C. Obstacle avoidance policies for cluster space control of nonholonomic multirobot systems. IEEE/ASME Transactions on Mechatronics. New York City, NY, USA: IEEE; 2012;**17**(6):1068-1079

[37] Khatib O. A unified approach for motion and force control of robot manipulators: The operational space formulation. Robotics and Automation, IEEE Journal of. 1987;**3**(1):1068

[38] Kopeikin A, Ponda SS, How JP. Control of communication networks for teams of UAVs. In: Valavanis KP, Vachtsevanos GJ, editors. Handbook of Unmanned Aerial Vehicles. Springer Science+Business; 2015

[39] Agnew MS, Canto PD, Kitts CA, Li SQ. Cluster space control of aerial robots. In: IEEE/ASME International Conference on Advanced Intelligent Mechatronics; Montreal, Canada. New York City, NY, USA: IEEE; 2010

[40] Ayorkor Korsah G, Stentz A, Bernardine Dias M. A comprehensive taxonomy for multi-robot task allocation. The International Journal of Robotics Research. 2013;**32**(12): 1495-1512

[41] Ponda SS, Johnson LB, Kopeikin AN, How JP. Distributed planning strategies to ensure network connectivity for dynamic heterogeneous teams. IEEE Journal on Selected Areas in Communications. June 2012;**30**(5): 861-869

Chapter 4

Controlling a Fleet of Autonomous LHD Vehicles in Mining Operation

Alexander Ferrein, Gjorgji Nikolovski, Nicolas Limpert, Michael Reke, Stefan Schiffer and Ingrid Scholl

Abstract

In this chapter, we report on our activities to create and maintain a fleet of autonomous load haul dump (LHD) vehicles for mining operations. The ever increasing demand for sustainable solutions and economic pressure causes innovation in the mining industry just like in any other branch. In this chapter, we present our approach to create a fleet of autonomous special purpose vehicles and to control these vehicles in mining operations. After an initial exploration of the site we deploy the fleet. Every vehicle is running an instance of our ROS 2-based architecture. The fleet is then controlled with a dedicated planning module. We also use continuous environment monitoring to implement a life-long mapping approach. In our experiments, we show that a combination of synthetic, augmented and real training data improves our classifier based on the deep learning network Yolo v5 to detect our vehicles, persons and navigation beacons. The classifier was successfully installed on the NVidia AGX-Drive platform, so that the abovementioned objects can be recognised during the dumper drive. The 3D poses of the detected beacons are assigned to lanelets and transferred to an existing map.

Keywords: autonomous vehicles, fleet control, planning, mining, computer vision, machine learning

1. Introduction

Along the lines of digitalisation and transforming industries towards Industry 4.0, also the mining industry as being a rather conservative industry is moving into this direction. For instance, in [1] the authors envision Mining 4.0 as a future concept of mining operations. The labour of a future mine worker will be smarter, more collaborative, more connected including augmented/virtual reality (AR/VR) technologies. In that sense, technology will also transform this sector into the Industry 4.0 direction. This has also an impact on the degree of automation deployed in future mines. In [2], mining is connected with key technologies such as intelligent systems, machine learning, and AR/VR. Additionally, there is a trend of the European raw materials industry towards changing from open pit mining to underground mining to reduce the environmental footprint of the mine. Digitalisation and automation are key technologies for further transforming mining operations into a decarbonised and more sustainable operations (see, e.g. [3–5]). This leads to hybrid mines, where parts of the mine are still open pit and parts are underground mines. This poses, in particular, additional challenges for the automation process. Whilst in the open pit part of the mine, methods from autonomous driving using GPS etc. can be deployed, in the underground part, methods from mobile robotics need to be used in an ever-changing environment with the need to continuously monitor and track the changes in the robots' maps.

Under certain limitations such as loaders following a prerecorded path, automated LHD vehicles are already commercially available [6]. However, the problem of fully automated guided load-haul dumpers remains a challenging problem. This results in a large body of related works that focus on the core problems like navigating and localising LHD vehicles. Some recent works are, for instance, [7], [8], or [9]. The former proposes topological navigation for underground haulage vehicles. The basic idea is that in underground mines with many tunnels, a highly precise localisation pose in the tunnel is not of importance, the LHD vehicle could localise at crossings or other important waypoints of a topological map. In our work, we also follow the idea of using topological maps, specifically in the Lanelet2 format [10], which is a common map format in autonomous driving.

In [8], a robust localisation system integrating cameras, LiDARs, and odometry information for underground LHD vehicles is proposed. The system is tested on mining datasets and shows good accuracy with mean errors below 1 meter. The methods deployed in this work are very similar to our approach. An interesting addition to common sensor cues is proposed in [9], where IMU data are used to record ground ripples as an additional localisation information. For measuring similarities and dissimilarities in the recorded data in order to recognise ground patterns, the dynamic time warping algorithm is deployed. The basic idea behind this approach is not to rely on visual landmarks such as visual tags to keep the extra infrastructure required in the mine as little as possible. On the other hand, extending the infrastructure in mines such as installing Wi-Fi at least in parts of the mine is no longer out of question. Some approaches make use of vehicle-to-vehicle (V2V) communication for establishing communication networks in underground mines [11]. The purpose of this work is to localise other vehicles underground deploying V2V communication. But localising other vehicles underground is not the only interest for mining operations as tracking mine workers underground is mandatory in many countries. Seguel et al. [12] overviews relevant positioning technologies to track mine workers underground. Finally, fleet control is one of the tasks that needs to be solved for automated haulage and artificial intelligence (AI) approaches are being deployed. Bnouachir et al. [13] overviews some approaches to intelligent fleet management in mining operations, whilst [14] proposes a real-time scheduling algorithm based on flow-achieving scheduling trees to overcome shortcoming of off-the-shelf software which often is based on myopic heuristics. In our case of fleet management, we make use of a planning approach which is based on hierarchical task networks [15].

Many related works concentrate on particular open problems in automating mining operations focusing mainly on localisation and navigation challenges or fleet-level planning focusing either on underground or open pit mines. Hybrid mines, which combine underground and open pit mine operations, pose a particular challenge for autonomous vehicles.

In this chapter, we report on the current state of affairs in our endeavour to automate hauling operations in hybrid mines. This work is based in part on our previous works [16–22]. In particular, we report on (1) the hardware setup of our fleet of robot vehicles including the LHD vehicles and a tracked exploration robot, (2) the overall ROS 2-based system architecture, (3) a model-based predictive control approach for controlling the articulated LHD vehicles, (4) an HTN-based approach to the tour planning of the fleet of vehicles.

The rest of this chapter is organised as follows. We present the hardware and sensor setup of the articulated LHD vehicles as well as of our exploration vehicle in Section 2 before we show the overall ROS 2-based software architecture (Section 3). Section 4 addresses the low-level navigation approach for controlling the articulated LHD vehicles. In Section 5, we outline the fleet and mission control system. Then, Section 6 reports on our approach to classifying drive ways in the mine and to mapping the changing mining environment. In Section 7, we show some experimental results. Then we conclude.

2. Hardware setup

In our mining automation projects, we are using two different platforms. One platform has been deployed for exploring and mapping the mine environment (Section 2.1), the other one is smaller-scale articulated dumping vehicle from Wacker Neuson that we turned into autonomous vehicle (Section 2.2).

2.1 Exploration vehicle

As mining environments are constantly changing, also the maps required for autonomous LHD vehicles need to be updated constantly. In our previous work [19, 23], we developed an exploration vehicle for mapping underground mines with dense 3D point clouds. As we can only give a brief overview of this work here, we refer to our previous work for further details.

The exploration vehicle is shown in **Figure 1a** and **b**. It is a skid-steered tracked robot based on a platform similar to a mini excavator but using a suspended undercarriage. It is equipped with a number of different sensors used for navigation, localisation, and mapping. The robot reaches speeds of up to 3 $\mathrm{m\ s^{-1}}$ and is controlled via the ROS [24] Movebase. For navigation, collision avoidance, and terrain classification, two Velodyne VLP-16 Puck LiDARS are mounted at the front. For mapping, we equipped the exploration vehicle with a custom-built rotating sensor platform shown in **Figure 1c** and **d**. It allows to acquire a (nearly) complete sphere around the vehicle. A Velodyne VLP-16 PUCK LiDAR with 16 scan lines, a vertical opening angle of 30°, and a horizontal range of 360° is mounted with a 14° inclination to the vertical axis. For short-range measurements, we additionally equipped the device with a 2D Hokuyo UTM-30LX-EW range scanner mounted in a 90° angle to the rotation plane. For teleoperation during mapping missions, an Allied Vision GT6600C high-resolution camera with a wide-angle lens is mounted at the front of the robot. As a safety feature, we mounted a FLIR A315 thermal camera at the front of the robot in order to be able to detect persons even when not sufficient light is available. Additionally, an IMU for providing the orientation of the platform with respect to the ground is mounted on the vehicle.

For registering the 3D point clouds from the scanning device, we deploy the Iterative Closest Point (ICP) method [25] based on the Point Cloud Library implementation [26]. To minimise the errors of pairwise registration of many point clouds, all point clouds are registered globally making use of the GraphSLAM [27] algorithm

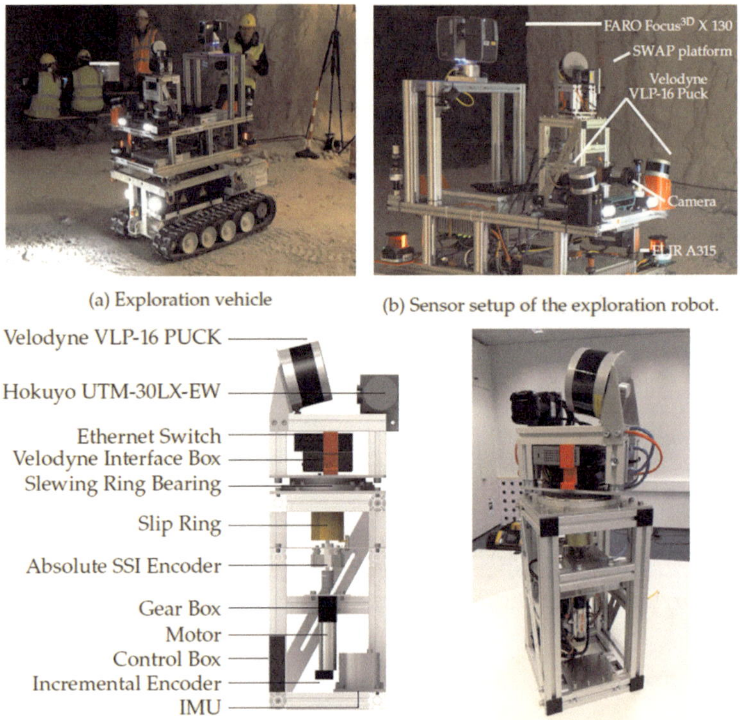

(a) Exploration vehicle

(b) Sensor setup of the exploration robot.

(c) Components of the platform.

(d) Photo of the platform.

Figure 1.
Exploration robot developed for mapping underground mining sites (from [23]).

after Lu and Milios [28]. It creates a graph of the connections between all point clouds with overlaps and minimises the alignment errors of all connections simultaneously. Finally, for an internal representation of the 3D map, Octomap [29] is used. This offers the advantage of being able to query the information in various resolutions and to map the distinctions that are important for navigation between free, occupied, and unknown cells.

We present an example map from an underground mine in Krölpa, Germany, in Section 7.

2.2 LHD vehicles

The base vehicle we use is an articulated haul-dumper from Wacker Neuson (Model 1501).[1] Our fleet consists of three of those vehicles. **Figure 2** shows an exemplary prototype. This chapter is based on a paper that has been previously published. We refer to [18] for further details.

The dumper is an off-the-shelf model that can handle loads of up to 1500 kg. The control of the brakes and the angle of the articulated joint of the vehicle is realised by a hydraulic system, which depends on the hydraulic pressure generated by a diesel engine. In order to automate the vehicle, we installed electric linear actuators with our

[1] https://www.wackerneuson.de/produkte/dumper/raddumper/raddumper-1501/

Figure 2.
Picture of one of the articulated haul-dumpers at the test site.

project partner Fritz Rensmann GmbH & Co. KG to control the brake and the throttle. Additionally, we attached a rotational servo motor to the steering axis and replaced the manual valves by electromagnetic valves to control the skip. We also installed hall sensors on each wheel to obtain information on the vehicle's state.

The control system comprises the different components shown in **Figure 3**. These parts are presented in the following. The software architecture of the vehicle will be discussed in the next section. As can be seen in the figure, each dumper can be controlled via our high-level control stack, where the tasks for the individual vehicles are generated by the overall mission planner for the whole mine to fulfil the daily production goals. The high-level control system is described in Section 5. Additionally, each vehicle can be controlled via remote control. In particular, this comes in handy when the vehicle needs to be steered onto a low loader trailer for transport.

Real-time Controller. The low-level real-time controller runs on a programmable logic controller by Beckhoff. To communicate with the motors and the PLC, we make use of a CAN interface. The PLC serves as a message filter, that prevents unsafe commands to be propagated to the motors, and as a kill-switch manager for controlling a number of kill-switches, which are installed on the vehicle including a radio kill-switch. The PLC runs a PID controller regulating the angle of the articulated joint and the engine speed by properly actuating the motor moving the steering axis and the motor opening and closing the throttle valve. For more detail, we refer to [30].

Compute Nodes. As described below (Section 4), we implemented a GPU-based model-predictive path-follower, a high-level control system, and various computer vision algorithms. All the software modules run on two dedicated compute devices.

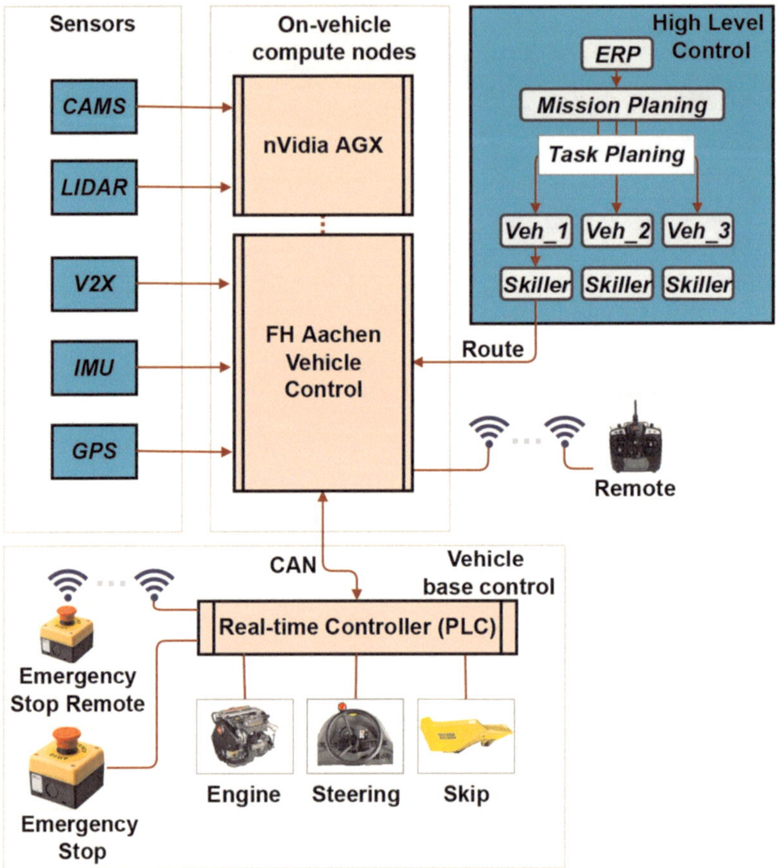

Figure 3.
System architecture of haul-dumper control system (From: [18]).

We make use of a Zotac ZBOX Magnus One including an NVIDIA GPU to run the path-follower, the high-level control, and semantic and life-long mapping algorithms. In addition, the computer vision tasks are deployed on an NVIDIA Drive AGX Xavier. All high-bandwidth sensors such as the cameras are interfaced to the AGX unit, which directly processes the data. Both compute devices run Linux as its operating system.

Sensors. In the open pit part of the hybrid mine, GPS localisation is possible. For this task, we deployed an OxTS RT3000v3 dGPS on one vehicle and equipped the two remaining vehicles with two OxTS xNav650 from Oxford Technical Solutions. The RT3000 shares its correction data via the xNAV650 devices. Each dGPS contains an IMU. Especially, in the underground part of the hybrid mine, where no GPS is available, we scanned the environment and made use of two VLP-16 Lidars and six cameras for localising and navigation. A mesh network spanning over multiple local stations in the testing area is used for vehicle interaction within the fleet and vehicle interaction with loading and unloading stations (V2X in **Figure 4**). Local access points are available on each of the vehicles that are connected to the mesh. For identification of the vehicles and direct addressing, we make use of ROS2 namespacing.

DOI: http://dx.doi.org/10.5772/intechopen.113044

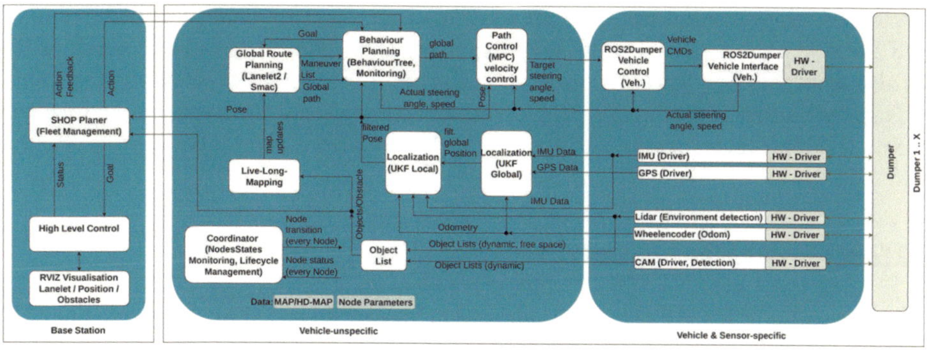

Figure 4.
Diagram of the software architecture (From: [18]).

3. Software architecture

In [21], we proposed a ROS 2-based architecture for self-driving cars. For our LHD vehicle, we adopted the architecture to fit the needs of the vehicle. Figure shows an overview of our architecture. The architecture also allows for using some common ROS packages such as Navigation2 [31] and robot_localisation [32]. In the following, we give an overview of the different parts of the architecture.

Centralised Mission Management Block. At the top-most level in the hierarchy of decision-making and control of the fleet is the high-level control block (leftmost block in **Figure 4**). In the LHD context, it means that a certain tonnage that should be hauled is defined and the high-level system is to find a plan for all the vehicles in the fleet to reach the tonnage. We deploy a SHOP3 [33] planning system for high-level planning. An additional fleet manager distributes the plans and dispatches the actions to the vehicles. Whilst the vehicles execute these actions, they continuously update their current positions and these of their locked resources. The world model gets information of the actual status of all agents and the hauled tonnage. If needed, in case of bigger differences to the original plan, a re-planning for the whole fleet is initiated.

Vehicle Unspecific Block. The vehicle unspecific section of the architecture consists of mid-tier functionalities, that are similar to three-tier robotics architectures with modules for localisation or path planning. A global route is planned via a free-space planner like smac or map-based planner using a Lanelet2 HD map. The vehicle follows this path utilising a path-following module, including a feedback loop [34, 35]. As path-following algorithm, we use a model-predictive controller (MPC) which uses a GPU-based grid-search on a set of predicted trajectories achievable by the vehicle's kinematic model. We introduce the details of the MPC in Section 4.

Vehicle Specific Block. The vehicle specific block (rightmost block in **Figure 4**) mostly consists of drivers and vehicle communication modules such as drivers for cameras, IMUs, GPS units, LiDARs, or the wheel encoders. For object detection, we have implemented an object detection with cameras using YOLOv5 to aid in the semantic and life-long mapping that is being presented in Section 6. We also implemented a modular way of integrating state-of-the-art deep neural networks for 3D object detection in point clouds. The detailed presentation of the latter is part of previous work presented in [36]. Our presented architecture is implemented using ROS 2 and deployed to each vehicle of the fleet. The underlying DDS network of ROS 2 is also used for the communication with the centralised high-level control system.

Next, we address the model-predictive controller used on the vehicle, before we discuss the high-level control software in Section 5.

4. Model-predictive control

In this section, we show the low-level control system based on a model-predictive control approach. We first introduce the kinematic model of the dumper, before we discuss the software implementation. Further details of the approach can be found in [17].

The kinematic model that has been used to model the haul-dumper in the MPC is described in [37] and shown in **Figure 5a**. This model has been chosen, as it models a centre-articulated platform and steering is done by changing the angle of the active joint. This model fits our haul-dumper well. The model is continuous and therefore the equations need to be discretised for our use case. The equations in the discrete form are as follows:

$$
\begin{aligned}
x_{t+1} &= x_t + \Delta_t{}^* (v\cos\psi) \\
y_{t+1} &= y_t + \Delta_t{}^* (v\sin\psi) \\
\psi_{t+1} &= \psi_t + \Delta_t{}^* \left(\frac{\sin\phi}{l_2 + l_1\cos\phi} v + \frac{l_2}{l_2 + l_1\cos\phi} \omega \right)
\end{aligned}
$$

The constant l_1 describes the length of the front part, l_2 the rear part of the vehicle, x and y are the current position values within the Cartesian coordinate system, and v the current velocity of the vehicle. The current steering angle is ϕ, ω denotes the angular velocity of the joint, and ψ is the heading of the front part within the Cartesian coordinate system as shown in **Figure 5a**. Δ_t represents the time interval between two control steps within the prediction process. These equations are used to predict the travel of the vehicle given the different inputs for each iteration of the optimisation.

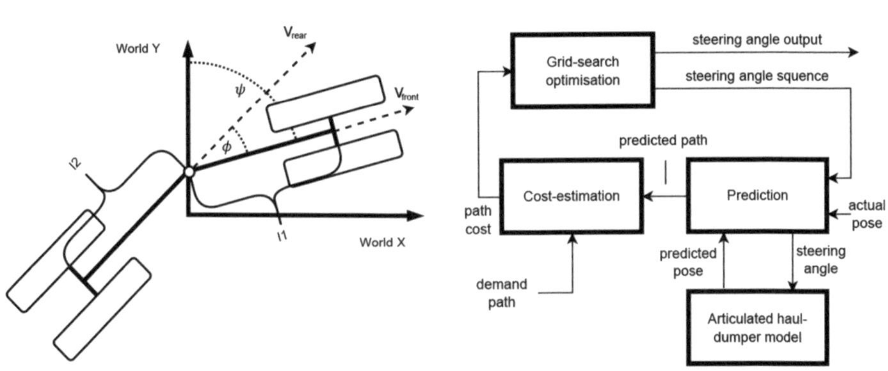

Figure 5.
Dumper kinematics and model-predictive control approach (From: [17]).

Figure 5b shows the MPC control cycle. Based on the actual pose and the articulated haul-dumper model (**Figure 5a**), a predicted path is calculated for a set of steering angle sequences. Subsequently, for each predicted path, the costs are estimated by three weights:

1. the *lateral error*, which weighs the difference in the vehicle's heading on the predicted path in contrast to the target trajectory;

2. the *orientation error*, which weighs the difference in the vehicle's lateral position on the predicted path in contrast to the target trajectory;

3. the *joint angle change*, which weighs the needed change of the joint angle from step to step.

Finally, in a grid-search optimisation, the output is determined as the first steering angle from the sequence of the steering angles for which the lowest cost was estimated.

The control result of this approach is very good, however, the calculation costs are very high. We therefore chose a GPU-based implementation of the controller based on our experiences described in [38]. The advantage of this approach is that the predicted paths and their individual costs can be calculated for all steering angle sequences in parallel in one step of the GPU. This leads to an improved control result at efficient calculation costs as we showed in [38].

Algorithm 1: MPC algorithm

Input : target trajectory $w_n = \{w_1, \ldots, w_n\}$, current vehicle pose p
Output: control command u
Data: n length of prediction horizon, m number of angle sequences

$\Phi_{n,m} \leftarrow$ *initial_set* ; /* set of steering angle sequences */
$S_{n,m} \leftarrow \emptyset$; /* set of predicted paths */
$c_m \leftarrow \emptyset$; /* set of cost evaluations */
$\tau_{lc} \leftarrow \emptyset$; /* trajectory index with least cost */

Perform grid search to find optimal trajectories:
for each *steering angle resolution* **do**
  Copy $\Phi_{n,m}$ and current vehicle pose p to GPU memory;
  for each *steering angle sequence* $\phi_n \in \Phi_{n,m}$ **calculate in parallel on GPU**
    $S_n \leftarrow Haul\_Dumper\_Model(p, \phi_n)$;
    $c \leftarrow Cost\_Function(S_n, w_n)$;
  end
  $\tau_{lc} \leftarrow Index\_Of(S_{n,m}(Min(c_m)))$;
  New steering angle sequence set for next iteration:
  $\Phi_{n,m} \leftarrow$ Sequence set around $\Phi_{n,m}(\tau_{lc})$;
end
Output first steering angle of optimal steering angle sequence:
$u \leftarrow \Phi_{n,m}(0, \tau_{lc})$;

The algorithm of the controller is shown in Algorithm 1, where the separation of GPU and CPU calculations is shown. To get an optimal trajectory resolution, multiple iterations are executed. The whole algorithm is calculated at a period of 20 ms. To adapt the original algorithm from [38] to the here-used long-haul-dumpers, we only had to change the kinematic model of vehicle and had to re-calibrate the control parameters (cf also to [17] for details).

5. High-level fleet planning

For larger-scale mining scenarios, several vehicles need to work together in a fleet. Such a fleet then needs to be coordinated for the work to be organised and distributed amongst the different vehicles in an optimised fashion. To realise the high-level fleet coordination system, we implemented a planning server embedding of SHOP 3 [33]. SHOP 3 is a domain-independent planning system based on ordered task decomposition. Ordered task decomposition is a modified version of hierarchical task network (HTN) planning [39] in which the planning order respects the actual order of execution of each task [33]. The interaction between the embedding sever and our vehicles is done by first communicating or loading a world model into the planner. Afterwards, a query for a day plan can be sent to the planner as a string, which the planner then processes. The query is a problem statement describing the resource allocation, available agents, and goals. As a result, the planner produces a string containing the day plan for all vehicles. This plan can then be distributed to the fleet. Each agent can then parse the plan and start executing the assigned actions.

As an example, **Figure 6** visualises a resource assignment. The blue circles in the figure represent resource sources. Above each source, we show the type of resource which can be loaded from that source and its quantity (rounded corners rectangles). The resource quantities of ores are modelled as infinite. The waste resource is constrained to 500 tons. The white circle represents a stockpiling point, which would be used to store resources if direct hauling to the unloading points would lead to congestion. In black, we colour the waste dumping points which can be used to unload waste from sources S1 and S2. The circles in yellow are fuel stations, that agents need to drive to fill their fuel tanks. Each agent in the problem statement is defined with a fuel state which is being used up when the agents transport resources. The arrows in the visualisation show possible direct load/unload relations. For example, the arrow from S1&2 to G13 means that the agent loads from S1 or S2 and unloads at the goal G13.

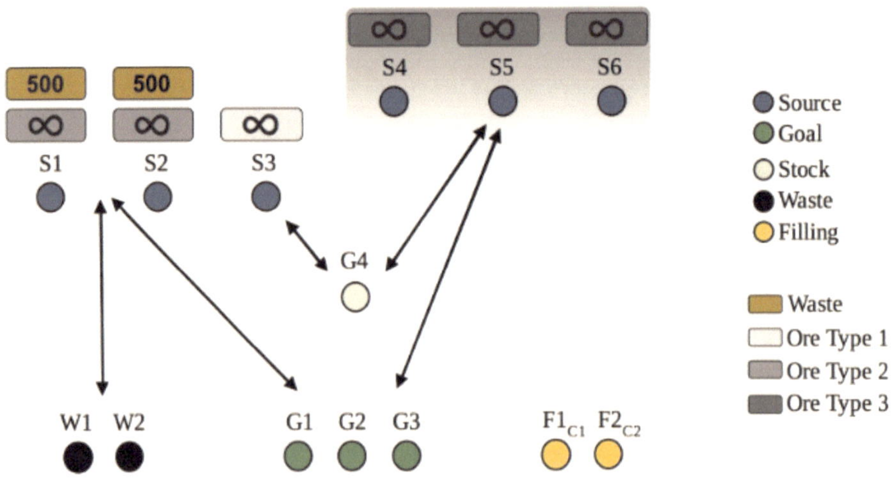

Figure 6.
Schematic of one of the problem statements evaluated in our research.

Module	KPIs					
Model Predictive Control	Mean lateral error in curves			Mean lateral error on straight section		
	0.9 (m)			0.1 (m)		
System Delays	Mean steering delay			Mean engine RPM delay		
	200 (ms)			100 (ms)		
Mission Planning (SHOP3)	Idle-time (min) of fleet after 8 h in simulation					
	Scenario Name	Number of vehicles simulated	Strategy 1	Strategy 2	Strategy 3	Strategy 4
	Simulation Scenario 1	75	25,782	26,096	26,046	25,938
	Simulation Scenario 2	40	2093	5343	3512	3084
	Overall hauled resources (t)					
	Scenario Name	Number of vehicles simulated	Strategy 1	Strategy 2	Strategy 3	Strategy 4
	Simulation Scenario 1	75	18,550	18,550	18,550	18,550
	Simulation Scenario 2	40	13,865	12,685	13,080	10,305
Object Detection	Model	mAP@0.5 Person	mAP@0.5 Wheel-dumper	mAP@0.5 Car	mAP@0.5 Beacon	Detection Frequency [hz]
	PointPillar (3D)[a]	0.34	0.4	0.4	—	20
	YOLOv5m (2D)	0.994	0.978	—	0.966	30

[a]*The results exclude performance for beacons, because the LiDAR sensor did not measure many points on the target beacons. Detection of beacons was, therefore, unreliable with the given hardware.*

Table 1.
Table on the quantified key performance indices for some of our modules we have researched by now (From: [18]).

Another problem statement we defined for the evaluation of the high-level control is less complex with only 5 sources with infinite resources and 2 goals. The meaning of sources and goals is the same as for **Figure 6**. The evaluation presented as an aggregation of key performance indicators in **Table 1** in Section 7 shows the results when different criteria are imposed on a plan. A plan can be made to take into consideration waiting times of the vehicles in a fleet, the time a vehicle needs to fulfil a load-unload cycle accumulated for a whole fleet, or idle time of the fleet as a whole. For comparison, we show the results for a plan that is created by generating plans randomly and selecting the first valid plan for a problem. This is the fourth strategy as presented in the evaluation. The plan that takes into consideration the waiting time of the agents is strategy 1. Strategy 2 takes the accumulated cycle time for all agents into account and tries to minimise it. The last strategy, strategy 3, minimises the time the vehicles in the fleet spend idling.

In our scenario, SHOP 3 instructs the navigation system to execute an action called `DRIVE_TO`. The action's main argument is an ID to a lanelet [10] in a map representing the routing graph. Lanelet is an HD map format which is an extension of OSM. With its integrated extensions for routing, it can support the execution of many tasks in navigation. Obstructions in the global map are represented within the routing graph and can be derived from an occupancy grid, which is implemented in the `costmap` ROS package. Given the goal, the action server runs a behaviour-tree [40] which computes the centre line of a route calculated by the `lanelet_planner`. This centre line is referenced to a world-fixed frame (i.e. the entrance of the mine) which is in turn passed on to our MPC to follow the path. Regular actuation commands are calculated by the MPC by calling its respective action server from within the behaviour-tree at a certain frequency.

6. Life-long mapping

With life-long mapping, we refer to the concept of mapping the environment and ensuring that changes in the environment can be detected and processed into a map update. To achieve this, we identified two approaches. First, one can use representations of a traffic network as maps and use static delimiting objects in the environment as markers that one can infer a change in the environment. Secondly, one can use shape representations of the environment (such as a point cloud) to extract a traffic network representation of the environment from the shape. To accommodate for changes in the environment, one can simply repeat the process of generating the traffic network from the environment.

For the first approach, three major components are needed: the object detection and classification of static and dynamic objects, the localisation of static object in a unified coordinate frame, and the integration of the information from the 2 previous components into an HD map. In the following, we explain this in more detail:

Object Detection. The artificial intelligence framework known as YOLOv5 was used for object recognition. The focus was on the recognition of humans, navigation aids such as beacons marking the way, and different types of vehicles commonly used in mining. To facilitate the recognition of previously unknown objects using this YOLOv5 mechanism, a three-part training methodology was devised:

Firstly, a solution using synthetic data was developed to automate the annotation process for training deep learning networks. By creating a realistic 3D mining world with Unreal Engine and capturing annotated images from virtual cameras on dumpers, see **Figure** 7, these synthetic datasets serve as cost-effective training data for the YOLOv5 network, addressing the challenges of manual annotation for new objects and environments.

As second stage, synthetically generated data for neural network training often lacks impurities found in real-world images, such as noise or blur. To address this, we developed an image augmentation tool. This tool introduces variations into the training images by adding noise, modifying lighting, saturation, image resolution, and horizontal alignment. By incorporating these variations, the synthetic training data becomes more realistic and better aligns with real-world conditions. Thirdly, the training data is supplemented by data from real journeys which are manually annotated and augmented.

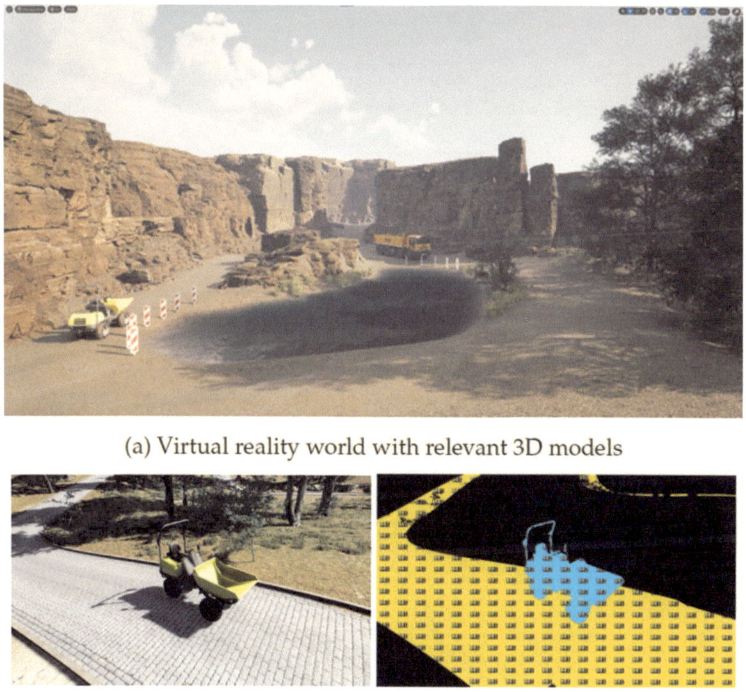

(a) Virtual reality world with relevant 3D models

(b) Simulated image (c) Labeled image

Figure 7.
Virtual reality world with 3D models created with unreal engine 5 to get annotated training data from simulated driving.

The classifier, trained with a blend of synthetic and real images over 299 epochs, attains 0.9835 mAP@0.5 and 0.9836 mAP@0.95 with YOLOv5m and effectively detects partially obscured objects. After converting from PyTorch to ONNX, then to TensorRT, it is deployed on Nvidia AGX Drive, with an inference time near 35 milliseconds.

Lane detection. Beacons serve as static boundary markers for navigable roads. We calculate the 3D position of these beacons by projecting the midpoint of the lower boundary box onto a ground plane, considering the camera's extrinsic parameters relative to the vehicle centre point. Finally, we derive the global UTM coordinates from the estimated vehicle-relative positions.

Boundary Matching and Map Correction. In order to correlate the positions of the beacons with a lane boundary, a filtration procedure is implemented. A binning filter reduces measurement noise by averaging close positions. A logical constraint filter ensures feasible interpolation of the lane boundary. Positions are then re-indexed based on their medial distance from the lane boundary segments. The updated lane boundary state is shown in **Figures 8** and **9**.

LiDAR Map To HD Map. Our approach for generating an HD map from a LiDAR map involves segmenting the navigable ground segment from the point cloud, calculating the concave hull, creating a Voronoi graph [41], finding the longest chain of vertices in the graph, smoothing the vertices, and converting the trajectory into a lane. The process includes steps such as elevation map creation, filtering of sample points based on a simple morphological filter from [42],

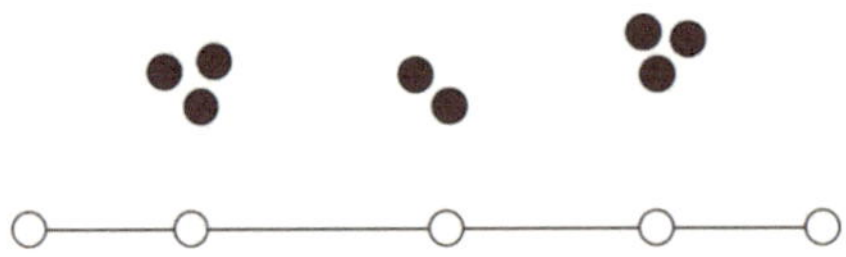

(a) Initial state of the estimate for the beacon positions (circles with infill) next to the lane boundary points (circles without infill).

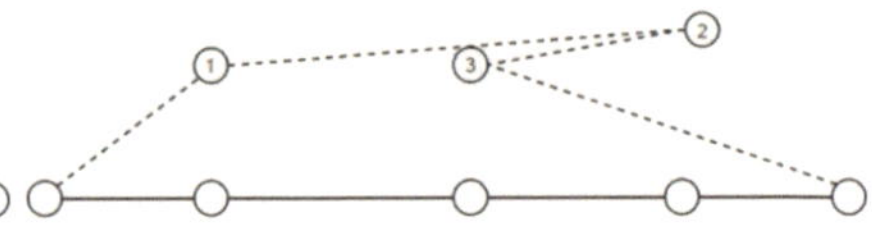

(b) State after filtering the positions of the beacons. Dotted line denotes the proposition for the new lane boundary at this point.

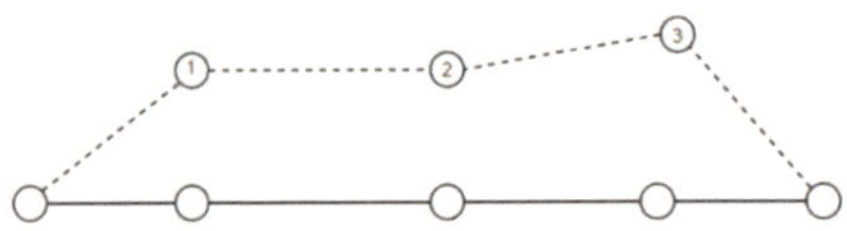

(c) State after re-indexing the beacon-position groups.

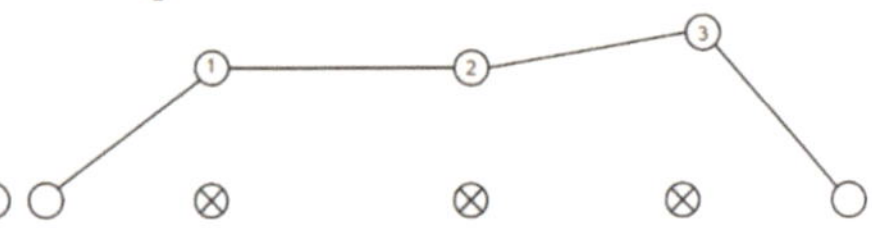

(d) In the final stage, outdated lane boundary points are discarded to form a new lane boundary.

Figure 8.
State of the lane boundary throughout the four stages of the lane boundary adjustment process.

(a) Snapshot of the 3D pose estimation. In blue one can see the bounding box of the object.

(b) Representation of untracked and unfiltered 3D-Poses rendered over an image of the lane.

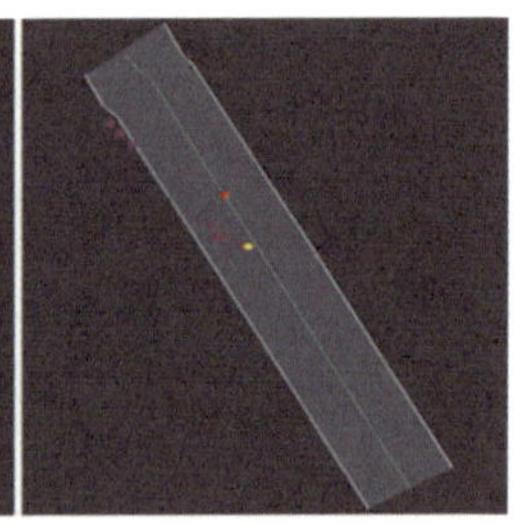

(c) Representation of filtered/grouped 3D-Poses rendered over an image of the lane.

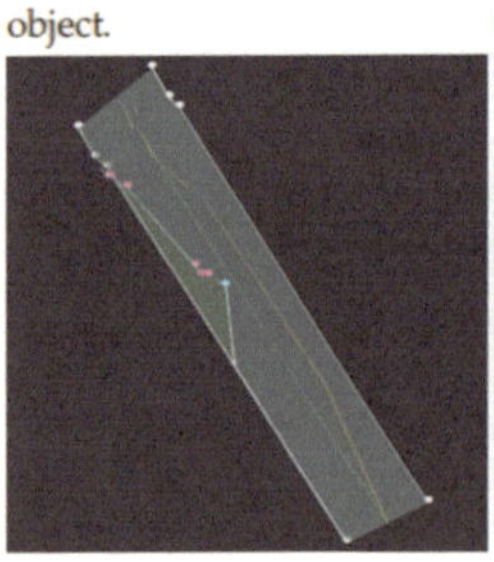

(d) Update proposal rendered on top of image of the lane with changed centre-line.

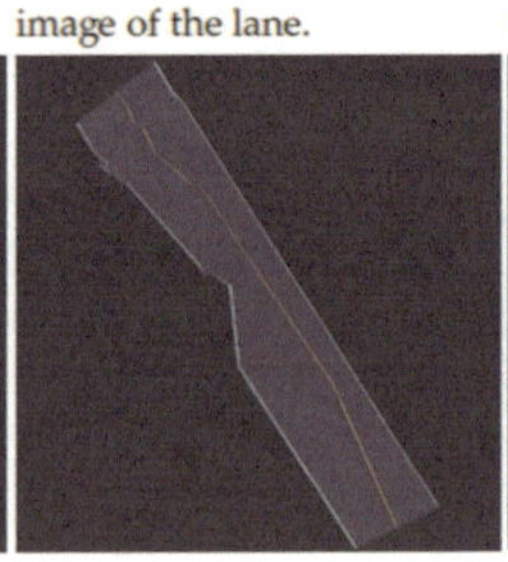

(e) Image of the updated lane.

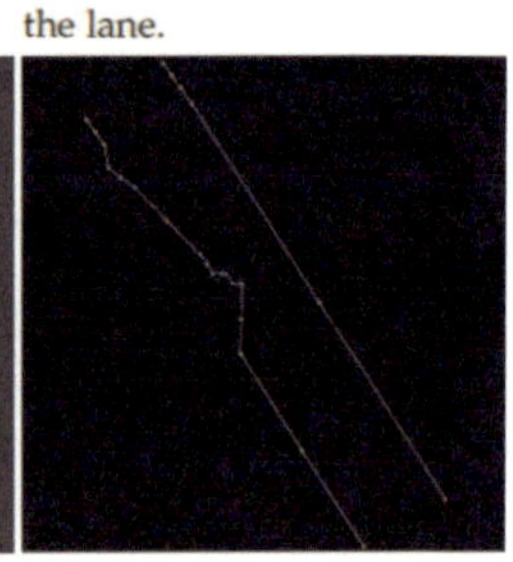

(f) Updated and valid lane rendered with the editing tool JOSM.

Figure 9.
Visualisation of the update process and its result.

calculating a spanning polygon, applying Delaunay triangulation, and constructing a lane with a consistent width. **Figures 10** and **11** show a visualisation of each step.

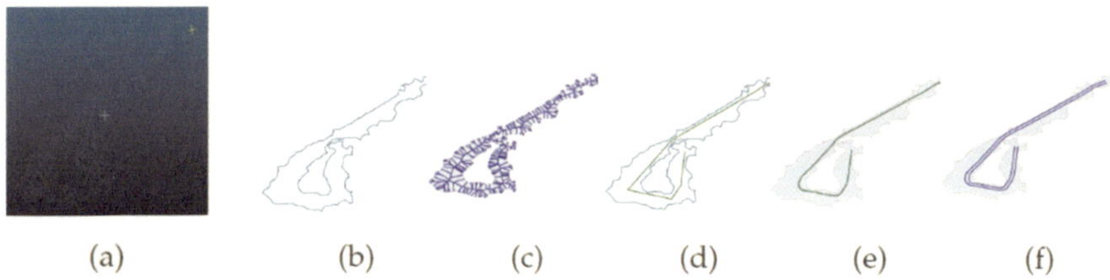

Figure 10.
Processed point cloud map after each step of the conversion from point cloud map to HD map.

Figure 11.
Visualisation of the integration scenario, which was presented at a demonstration day.

7. Real-world experiments

In this section, we show some experimental results of our work.

3D Mapping. Regarding underground mapping with the exploration vehicle described in Section 2.1, we show maps from the MAXIT underground anhydrite mine in Krölpa, Germany in **Figure 12**. The map was recorded by exploration vehicle in a stop-and-scan fashion taking a full spherical 3D scan every 10 m. The recording of the map took place in a teleoperated manner, the point cloud data and the odometry data were stored in a rosbag file and were processed offline following the procedure described in detail in [19, 20]. **Figure 12a** shows a 2D occupancy grid of the mapped part of the mine. **Figure 12b** visualises the point clouds themselves and **Figure 12c** a 3D Octomap from the point clouds. Whilst the results show that precise 3D maps of the underground mining environment can be produced with the exploration vehicle, one has to admit, on the downside, that the process is not fully automated and human expertise is required to avoid errors in registering the different point clouds into a consistent map. The overall map size of the map is about 800 m.

Vehicle Automation. The haul-dumper's hardware setup has proved reliable throughout our project, with hundreds of testing hours and no system-wide failures. The most frequent issue involves power supply to the vehicle's components. The installed PLC and sluggish system caused by the hydraulic mechanism lead to an average latency of 200 ms from command issuance to actuators. There is also latency

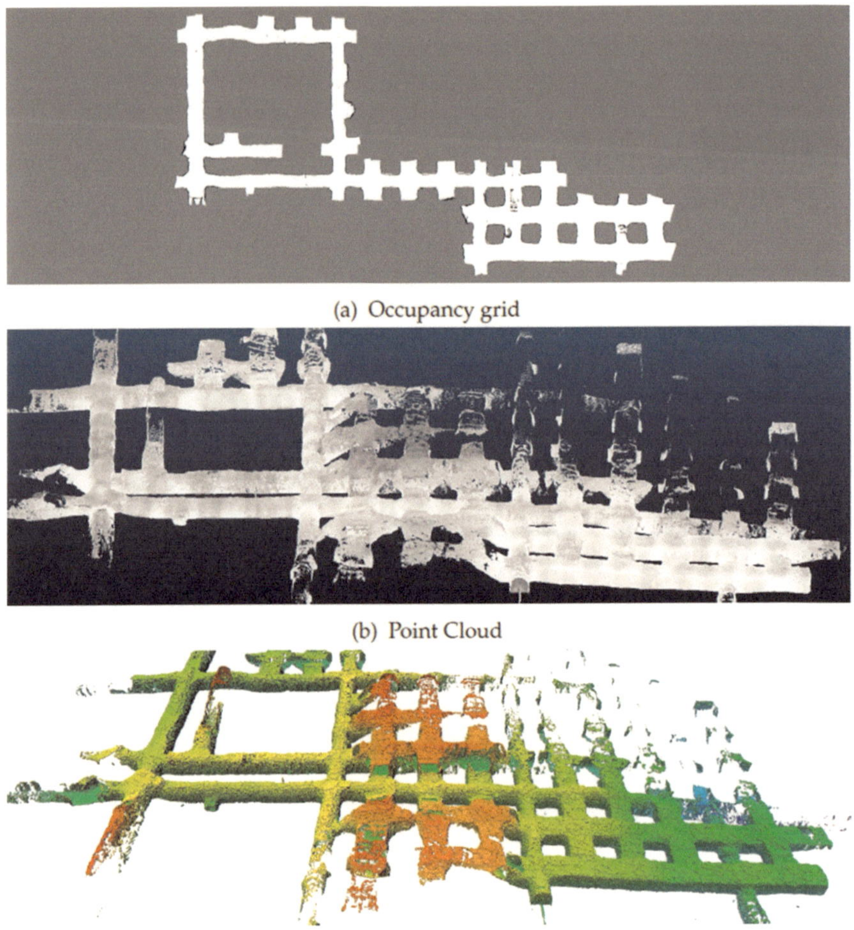

(a) Occupancy grid

(b) Point Cloud

(c) Octree

Figure 12.
Parts of the MAXIT anhydride mine in Krölpa, Germany.

in emergency stops, taking 1.5 seconds to fully halt after triggering the switch. The path-following feature shows a mean lateral deviation of 0.1 m on straight sections and 0.9 m on curves.

Fleet-level Planning. Fleet-level planning was tested through coarse simulations that mimic mining processes and through integration in the lower modules of the architecture. Two simulation scenarios were created with different levels of complexity: one with simple targets and resources, the other emulating a hybrid mine with multiple resource types. In each scenario, four strategies were analysed, each examining a different heuristic: Minimising the idle time of all vehicles in the fleet, minimising the average duration of loading and unloading the dump truck, minimising the idle time of the dump truck without moving loads, and maximising the haulage mass from randomly generated plans. Idle time and total haulage mass were observed in an 8-h simulation. The results are shown in **Table 1**.

Long-term Mapping. We carried out a functional test on live map correction by manually altering the map and monitoring the global route planner and vehicle response. This often involved tweaking the lane sections leading to the loading station, as demonstrated in the integration scenario **Figure 11**.

Integration Scenario. In an integration test, we utilised loading and unloading stations to handle pick-up and unload payloads. The unloading heap has two access points, whereas the loading station has only one. There are two bidirectional lanes to the unloading sites and a single bidirectional lane to the loading site.

The daily transport schedule of fleet management includes repeated loading and unloading operations for the vehicle fleet, with only one vehicle having access to the loading point. The high-level controller coordinates resource access to avoid conflicts. Resource blocking is implemented by the operations scheduling system. The first vehicle to reach the fork in the middle of the traffic network whilst the loading resource is free secures the lock and blocks the lane to the resource. Following vehicles must wait for fleet management to release the resource. This mechanism proved robust in an 8-h operation and successfully implemented the daily schedule.

8. Conclusion

In this chapter, we present our results from automating load-haul-dump (LHD) operations in hybrid mines. Hybrid mining operations are mines where mining in part is done in an open pit fashion, and parts are underground. As for the autonomous vehicles deployed in such a mine, it means that the vehicles cannot simply rely on GPS data for localising themselves over ground, but also need classical approaches to mapping the environment. We reported on our exploration vehicle that is equipped with a rotating LiDAR scanner to produce detailed 3D point clouds which then are integrated into Octree maps that the LHD vehicles could use for localising themselves in the mine.

The haul-dumpers that we deploy are modified off-the-shelf articulated dumpers which were turned into autonomous vehicle. The vehicles are equipped with GPS, LiDAR, and camera sensors. The software architecture is based on ROS2 and was also deployed in similar projects related to autonomous driving including a modified version of a model-predictive control (MPC) algorithm. It had to be adapted to the articulated kinematics of the dumper. Further, to control a fleet of LHD vehicles (three in our case), we made use of the HTN-based planner SHOP3, which generates a global mission plan for the vehicles based on the required haulage capacities. Each vehicle is following this global plan for generating their local missions. As another important contribution, we introduced Lanelet2-based maps for navigating the LHD vehicles, including dedicated drive ways and right of way rules. Lanelet maps are commonly used for self-driving cars and not so much in robotics applications. Using this approach facilitates the coordination of a fleet of vehicles to a great extend. Finally, we showed our approach to object and drive-way detection including the automatic generation of the aforementioned Lanelet drive ways.

The presented approach has been tested in real-world scenarios underground and in open pit mines under controlled conditions. We showed how mapping successfully took place with our exploration vehicle in the Krölpa mine in Germany. The fleet of dumpers was tested in an open gravel pit in Buir, Germany. We could show that the low-level control such as MPC works well also on the dumpers with their articulated kinematics and how the fleet of dumpers could be coordinated on a mission level. Whilst the tests and experimental results show that the overall approach is working, next steps would be to deploy this work in real mining operations under the realistic hard mining environment conditions with limited communication means.

Acknowledgements

We gratefully acknowledge the contributions of (in alphabetical order) N. Adam, A. Braining, D. Bulla, M. Claer, B. Decker, J. D. Eichenbaum, F. Engels, J. Grabe, K. Kramer, K. Krückel, M. Lubos, A. Lyrmann, T. Neumann, Y. Otten, J.C. Richter, D. Scholl, P. Shah, P. Stricker, and S. Weyer. The work presented in this chapter was funded with grants from the German Federal Ministry of Education and Research (BMBF) under grant numbers 033R126C, 033R126CN.

Author details

Alexander Ferrein*, Gjorgji Nikolovski, Nicolas Limpert, Michael Reke, Stefan Schiffer and Ingrid Scholl
Mobile Autonomous Systems and Cognitive Robotics Institute, FH Aachen - University of Applied Sciences, Aachen, Germany

*Address all correspondence to: ferrein@fh-aachen.de

References

[1] Lööw J, Abrahamsson L, Johansson J. Mining 4.0—The impact of new technology from a work place perspective. Mining, Metallurgy & Exploration. 2019;**36**:701-707

[2] Faz-Mendoza A, Gamboa-Rosales NK, Medina-Rodríguez C, Casas-Valadez MA, Castorena-Robles A, López-Robles JR. Intelligent processes in the context of mining 4.0: Trends, research challenges and opportunities. In: 2020 International Conference on Decision Aid Sciences and Application (DASA). IEEE; 2020. pp. 480-484

[3] Batterham R. The mine of the future – Even more sustainable. Minerals Engineering, Sustainable Minerals. 2017; **107**:2-7

[4] Sánchez F, Hartlieb P. Innovation in the mining industry: Technological trends and a case study of the challenges of disruptive innovation. Mining, Metallurgy & Exploration. 2020;**37**(5): 1385-1399

[5] Clausen E, Sörensen A. Required and desired: Breakthroughs for future-proofing mineral and metal extraction. Mineral Economics. 2022;**35**(3):521-537

[6] Paraszczak J, Gustafson A, Schunnesson H. Technical and operational aspects of autonomous lhd application in metal mines. International Journal of Mining, Reclamation and Environment. 2015;**29**(5):391-403

[7] Mascaró M, Parra-Tsunekawa I, Tampier C, Ruiz-del Solar J. Topological navigation and localization in tunnels—Application to autonomous load-haul-dump vehicles operating in underground mines. Applied Sciences. 2021;**11**(14):6547

[8] Jacobson A, Zeng F, Smith D, Boswell N, Peynot T, Milford M. What localizes beneath: A metric multisensor localization and mapping system for autonomous underground mining vehicles. Journal of Field Robotics. 2021; **38**(1):5-27

[9] Stefaniak P, Jachnik B, Koperska W, Skoczylas A. Localization of LHD machines in underground conditions using IMU sensors and DTW algorithm. Applied Sciences. 2021;**11**(15):6751. DOI: 10.3390/app11156751

[10] Poggenhans F, Pauls J-H, Janosovits J, Orf S, Naumann M, Kuhnt F, et al. Lanelet2: A high-definition map framework for the future of automated driving. In: 2018 21st International Conference on Intelligent Transportation Systems (ITSC). IEEE; 2018. pp. 1672-1679

[11] Dong L, Sun D, Han G, Li X, Hu Q, Shu L. Velocity-free localization of autonomous driverless vehicles in underground intelligent mines. IEEE Transactions on Vehicular Technology. 2020;**69**(9):9292-9303

[12] Seguel F, Palacios-Játiva P, Azurdia-Meza CA, Krommenacker N, Charpentier P, Soto I. Underground mine positioning: A review. IEEE Sensors Journal. 2022;**22**(6):4755-4771

[13] Bnouachir H, Chergui M, Machkour N, Zegrari M, Chakir A, Deshayes L, et al. Intelligent fleet management system for open pit mine. International Journal of Advanced Computer Science and Applications. 2020;**11**(5):327-332

[14] Seiler KM, Palmer AW, Hill AJ. Flow-achieving online planning and dispatching for continuous transportation with autonomous

vehicles. IEEE Transactions on Automation Science and Engineering. 2020;**19**(1):457-472

[15] Erol K, Hendler JA, Nau DS. UMCP: A sound and complete procedure for hierarchical task-network planning. In: Proceedings of the Second International Conference on Artificial Intelligence Planning Systems (AIPS'94). AAAI Press; 1994. pp. 249-254

[16] Eichenbaum J, Nikolovski G, Mülhens L, Reke M, Ferrein A, Scholl I. Towards a lifelong mapping approach using lanelet 2 for autonomous open-pit mine operations. In: Proc. IEEE CASE 23. IEEE; 2023

[17] Nikolovski G, Limpert N, Nessau H, Reke M, Ferrein A. Model-predictive control with parallelised optimisation for the navigation of autonomous mining vehicles. In: 2023 IEEE Intelligent Vehicles Symposium, IV 2023. IEEE; 2023

[18] Ferrein A, Reke M, Scholl I, Decker B, Limpert N, Nikolovski G, et al. Towards a fleet of autonomous haul-dump vehicles in hybrid mines. In: ICAART (1). SCITEPRESS; 2023a. pp. 278-288

[19] Ferrein A, Scholl I, Neumann T, Krückel K, Schiffer S. A system for continuous underground site mapping and exploration. In: Reyhanoglu M, Cubber GD, editors. Unmanned Robotic Systems and Applications. London, UK, Rijeka: IntechOpen; 2019

[20] Donner R, Rabel M, Scholl I, Ferrein A, Donner M, Geier A, et al. Die Extraktion bergbaulich relevanter Merkmale aus 3D-Punktwolken eines untertagetauglichen mobilen multisensorsystems. Tagungsband Geomonitoring. 2019;**S**:91-110. DOI: 10.15488/4515

[21] Reke M, Peter D, Schulte-Tigges J, Schiffer S, Ferrein A, Walter T, et al. A self-driving car architecture in ROS2. In: 2020 International SAUPEC/RobMech/PRASA Conference. IEEE; 2020. pp. 1-6

[22] Macenski S, Foote T, Gerkey B, Lalancette C, Woodall W. Robot operating system 2: Design, architecture, and uses in the wild. Science Robotics. 2022;7(66):eabm6074

[23] Neumann T, Dülberg E, Schiffer S, Ferrein A. A rotating platform for swift acquisition of dense 3D point clouds. In: ICIRA 9834 of Lecture Notes in Computer Science. Cham: Springer; 2016. pp. 257-268

[24] Quigley M, Conley K, Gerkey B, Faust J, Foote T, Leibs J, et al. Ros: An open-source robot operating system. In: ICRA Workshop on Open Source Software. Vol. 3. Kobe, Japan; 2009. p. 5

[25] Besl P, McKay ND. A method for registration of 3-D shapes. IEEE Transactions on Pattern Analysis & Machine Intelligence. 1992;**14**(2):239-256

[26] Rusu RB, Cousins S. 3D is here: Point cloud library (PCL). In: 2011 IEEE International Conference on Robotics and Automation. IEEE; 2011. pp. 1-4

[27] Thrun S, Montemerlo M. The graph SLAM algorithm with applications to large-scale mapping of urban structures. The International Journal of Robotics Research. 2006;**25**(5–6):403-429

[28] Lu F, Milios E. Robot pose estimation in unknown environments by matching 2d range scans. Journal of Intelligent and Robotic Systems. 1997; **18**(3):249-275

[29] Hornung A, Wurm KM, Bennewitz M, Stachniss C, Burgard W. OctoMap: An efficient probabilistic 3d

mapping framework based on octrees. Autonomous Robots. 2013;**34**(3):189-206

[30] Sürken D. Programmierung einer Basisautomatisierung für eine Wacker Neuson 1501 Modellmulde [Thesis]. FH Aachen University of Applied Sciences in German; 2021

[31] Macenski S, Martin F, White R, Ginés Clavero J. The marathon 2: A navigation system. In: 2020 IEEE/RSJ International Conference on Intelligent Robots and Systems (IROS). IEEE; 2020

[32] Moore T, Stouch D. A generalized extended kalman filter implementation for the robot operating system. In: Proceedings of the 13th International Conference on Intelligent Autonomous Systems (IAS-13). Cham: Springer; 2014

[33] Goldman RP, Kuter U. Hierarchical task network planning in common lisp: The case of SHOP3. In: Proceedings of the 12th European Lisp Symposium (ELS), April 1-2, 2019, Genova, Italy. Zenodo; 2019. pp. 73-80. DOI: 10.5281/zenodo.2633324

[34] Pivtoraiko M, Knepper RA, Kelly A. Optimal, Smooth, Nonholonomic Mobile Robot Motion Planning in State Lattices. Pittsburgh, PA: Robotics Institute, Carnegie Mellon University; 2007

[35] Limpert N, Schiffer S, Ferrein A. A local planner for Ackermann-driven vehicles in ROS SBPL. In: 2015 Pattern Recognition Association of South Africa and Robotics and Mechatronics International Conference (PRASA-RobMech). IEEE; 2015. pp. 172-177

[36] Nikolovski G, Reke M, Elsen I, Schiffer S. Machine learning based 3D object detection for navigation in unstructured environments. In: IEEE Intelligent Vehicles Symposium Workshops, IV 2021. IEEE; 2021. pp. 236-242

[37] Delrobaei M, McIsaac KA. Design and steering control of a center-articulated mobile robot module. Journal of Robotics. 2011;**2011**:621879

[38] Chajan E, Schulte-Tigges J, Reke M, Ferrein A, Matheis D, Walter T. Gpu based model-predictive path control for self-driving vehicles. In: 2021 IEEE Intelligent Vehicles Symposium (IV). IEEE; 2021. pp. 1243-1248

[39] Dvorak F, Barták R, Bit-Monnot A, Ingrand F, Ghallab M. Planning and acting with temporal and hierarchical decomposition models. In: 2014 IEEE 26th International Conference on Tools with Artificial Intelligence. IEEE; 2014. pp. 115-121

[40] Ghzouli R, Berger T, Johnsen EB, Dragule S, Wasowski A. Behavior trees in action: A study of robotics applications. In: Proceedings of the 13th ACM SIGPLAN International Conference on Software Language Engineering. New York, NY, United States: Association for Computing Machinery; 2020. pp. 196-209

[41] Bhattacharya P, Gavrilova ML. Voronoi diagram in optimal path planning. In: 4th International Symposium on Voronoi Diagrams in Science and Engineering (ISVD 2007). IEEE; 2007. pp. 38-47

[42] Pingel TJ, Clarke KC, McBride WA. An improved simple morphological filter for the terrain classification of airborne lidar data. ISPRS Journal of Photogrammetry and Remote Sensing. 2013;**77**:21-30

Chapter 5

Geometric Control of Robotic Systems

Anis Bousclet

Abstract

In this chapter, we'll present geometrical methods to control robotic systems, we'll start by presenting the classical approach and its benefits, as well as its limits, and we will explain how the geometric approach allow us to get rid of these limits.

Keywords: robotic systems, Riemannian geometry, geometric control, optimal control, regulation under constraints

1. Introduction

Robotic systems are powerful tools that allow us to gain time, money and enery, they had revolutionized several fields as industry, medicine, agriculture and many other fields.

The main difficulty caused by robotic systems is the nonlinearity of the dynamics. The classical approach modelise the configuration space by an Euclidean space, and applies the principle of least action to derive the equation of motion, and applies Lyapunov and Lasalle's techniques to control the robot [1, 2].

However, this elegent approach do not fit well with regulation under constraints problem or optimal regulation, the goal here is to illustrate these limits, and to show how the geometric method allow us to get ride of these limits.

2. The classical approach and its limits

The configuration space of the robot is $\mathbb{R}^n$ where n is the number of degrees of liberty (DOF) of the robot, the principle of least action gives the dynamics

$$G(q)q'' + \frac{1}{2}G(q)' + S(q,q')q' + g(q) = \tau. \quad (1)$$

This allow us to ensure locally the regulation to a configuration $q^* \in \mathbb{R}^n$ with the control

$$\tau(q,q') = g(q) - K_p(q - q') - K_v q' \quad (2)$$

which is a PD controller.

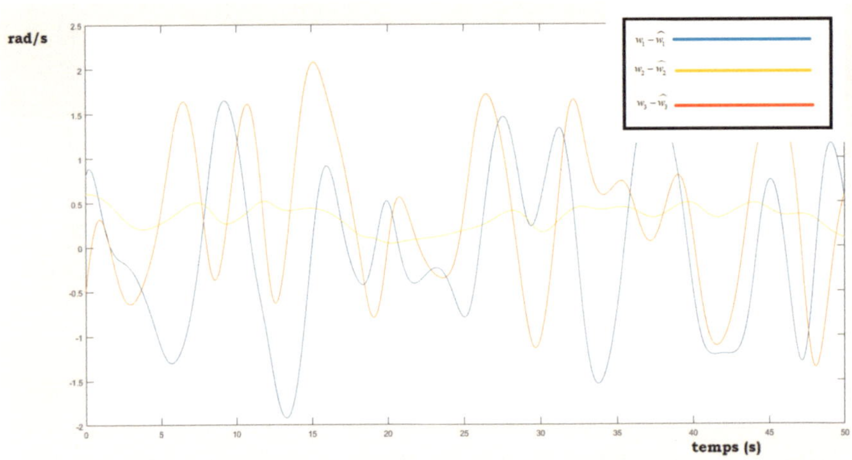

Figure 1.
Insufficiency of Luenberger observer.

The classical approach with all its benefits [1, 2], do not fit well with the problem of regulation under constraints or optimal regulation.

For the optimal control problem, the general formula which gives the angular velocity in function of the configuration of a Rigid body that minimizes the cost [3]

$$J(\Omega) = \int_0^\infty \left[tr\left(I - R_d^T R\right) + \frac{1}{2} \|\Omega\|^2 \right] dt \tag{3}$$

for the kinematic $R' = R\Omega$ and ensure regulation $R \to R_d$

$$\Omega^*(R) = -\frac{R_d^T R - R^T R_d}{\sqrt{1 + tr\left(R_d^T R\right)}} \tag{4}$$

which has not the form of a proportional controller, so this prevents to get an optimal PD regulator for robotic systems.

For the regulation under constaints problem (tool of the robot must stay in a security surface S), the set of admissibles configurations is $N = x^{-1}(S)$ where x is the function which gives the position of the tool in function of the configuration. The set N has no reason to be a linear subspace of R since x is a nonlinear function, the set N will be a curved space in $\mathbb{R}^n$ and we arrive to a formalism naturally adapted to the resolution of the problem of regulation under constraints.

Sensitivity with respect to initial conditions remains one of the main problems in conceptions of efficients observers.

For example, for a rigid body which dynamics are[1,2]

$$R' = R\Omega \tag{5}$$

$$\omega' = I^{-1}(I\omega \times \omega) - 2I^{-1}[Skew(R)]^{\times} \tag{6}$$

---

[1] $[\Omega]^{\times}$ is the unique vector ω such that $\omega \times x = \Omega.x$, and j is its inverse.

[2] $Skew(R) = \frac{R - R^T}{2}$.

where $I = [2, 2, 8]$ and $R(0) = \exp([0.82, 0.21, 0.73]^\times)$ $\omega(0) = [0.8, 0.5, 0.6]$, the Luenberger observer [4] is

$$\hat{R}' = \hat{R}\left(\hat{\Omega} - 2j\left(I^{-1}\left[Skew\left(R^T\hat{R}\right)\right]^\times\right)\right) \tag{7}$$

$$\hat{\omega}' = I^{-1}(I\hat{\omega} \times \hat{\omega}) - 2I^{-1}[Skew(R)]^\times - 2I^{-1}\left[Skew\left(R^T\hat{R}\right)\right]^\times \tag{8}$$

where $\hat{R}(0) = \exp([0.8, 0.5, 0.6]^\times)$ and $\hat{\omega}(0) = [0, 0, 0]$, the simulations shows that the Luennberger observer is not efficient when real and observed configuration and velocity are lightly different (**Figure 1**).

3. The geometric formalism

The idea of the geometric formalism is to pass the complexity of the equation to the configuration space, thus in the geometric formalism, the configuration space is not an Euclidean space, but a curved space that we call manifold.

Manifolds [5] are objects that looks like Euclidean space locally, we can always reduce the local study of manifolds to computations in $\mathbb{R}^n$, it is exactly what we do when we parametrize the configuration of a rigid body with Euler angles.

The configuration space of a rigid body [6] which has a fixed point is the space of Euclidean rotations $SO(3)$, when all its points moves freely the configuration space becomes $SO(3) \times \mathbb{R}^3$, the possible translations of a point cover $\mathbb{R}^3$.

A robot is an assembly of s rigid bodies, which are related from there extremeties by articulations, those links are called holonomic constraints.

The configuration space of the robot is M a submanifold of dimension n of $\left(SO(3) \times \mathbb{R}^3\right)^s$, the holonomic constraints are related to the local coordinates by the following result.

Proposition 1 *Let M be a subset of* $\left(SO(3) \times \mathbb{R}^3\right)^s$, *there is equivalence between*

- *Holonomic constraints: for each* $q \in M$, *there exists* V_q *a neighborhood of* q *in* $\left(SO(3) \times \mathbb{R}^3\right)^s$ *and a submersion*[3] $f : V_q \to \mathbb{R}^{6s-r}$ *such that* $M \cap V_q = f^{-1}(0)$.

- *Local coordinates: for each* q *in M, there exists a neighborhood* V_q *of* q *and an immersion*[4] $\theta : 0 \in U \to V_q \cap M$ *is a homeomorphism.*[5]

M is said to be a manifold of dimension n, the number n is the degree of freedom of the robot, and the number r is the number of constraints, we have the well known formula $n = 6s - r$.

Once we know that we are dealing with curved objects, it may seems more complicated to do computations. Thanks to the tangent space T_qM, which is the set of virtual velocities $\dot{\gamma}(0)$ of curves $\gamma(t)$ that pass throught $\gamma(0) = q \in M$ and stay in $\gamma(t) \in M$, this space is linear and of dimension n (**Figure 2**).

---

[3] A submersion is a smooth function for which the differential is everywhere onto.

[4] An immersion is a smooth function for which the differential is everywhere one to one.

[5] Invertible continuous map with continuous inverse.

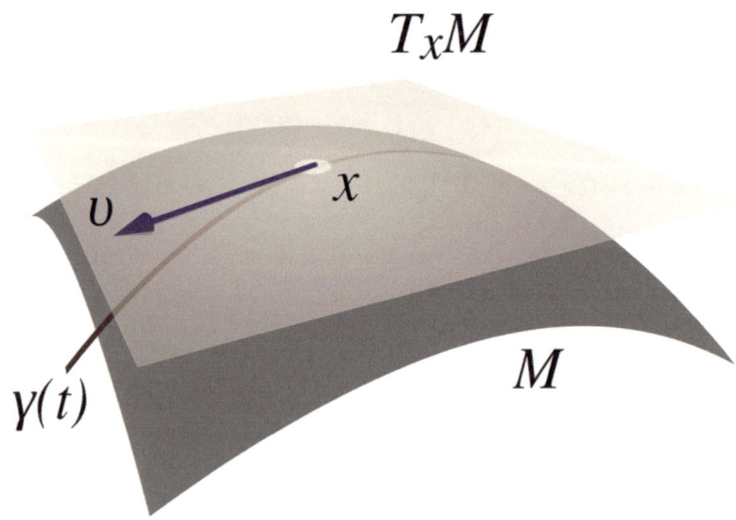

Figure 2.
Tangent space.

When we do parametrization with holonomic constraints, the tangent space is $T_qM = Ker\left(df_q\right)$, if we parametrize our manifold with local coordinates, then $T_qM = Range(d\theta_0)$, the basis adapted to the coordinates is $\partial_i = d\theta(e_i)$ where e_i is the canonical basis of $\mathbb{R}^n$.

The tangent linear map between two manifolds $\phi : M \to N$ is the linear map

$$T_q\phi : T_qM \to T_{\phi(q)}N \tag{9}$$

defined by

$$T_q\phi(v) = (\phi\circ\gamma)'(0) \tag{10}$$

where $\gamma(0) = q$, $\gamma(t) \in M$ and $\gamma'(0) = v$ (**Figure 3**).

A vector field is the data of a tangent vector in T_qM for each $q \in M$ which a smooth way, the set of vector fields is $\Gamma(TM)$.

To each vector field corresponds a motion of diffeomorphisms[6] by solving the following ODE (**Figure 4**)

$$\partial_t\phi_t^X(q) = X\left(\phi_t^X(q)\right) \tag{11}$$

$$\phi_0^X(q) = q. \tag{12}$$

When we have two vector fields X, Y, a natural question is to ask when we have

$$\phi_t^X\circ\phi_s^Y = \phi_s^Y\circ\phi_t^X \tag{13}$$

the obstruction of the commutation between the flows is measured by the Lie bracket $[X, Y]$ which is defined by

---

[6] Diffeomorphism is an invertible smooth mapping with smooth inverse

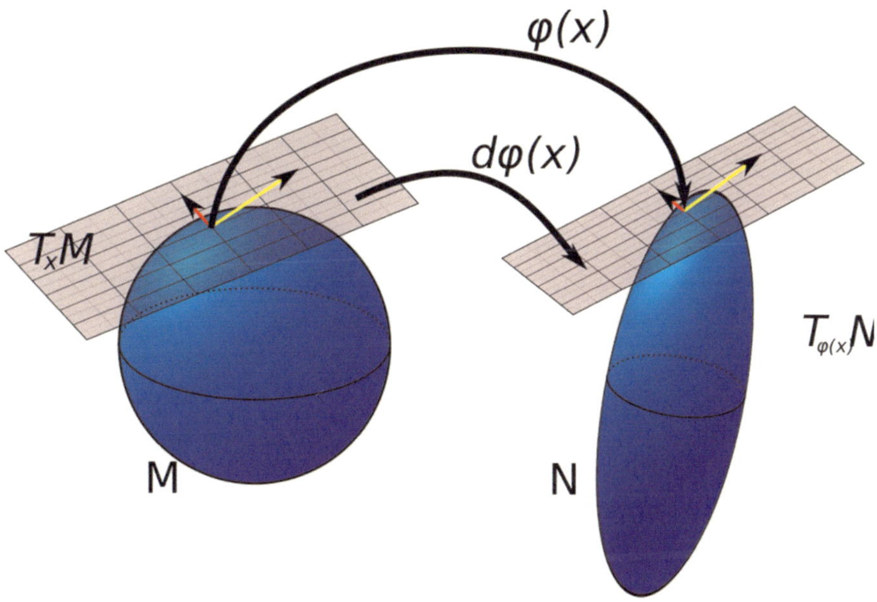

Figure 3.
Linear tangent map.

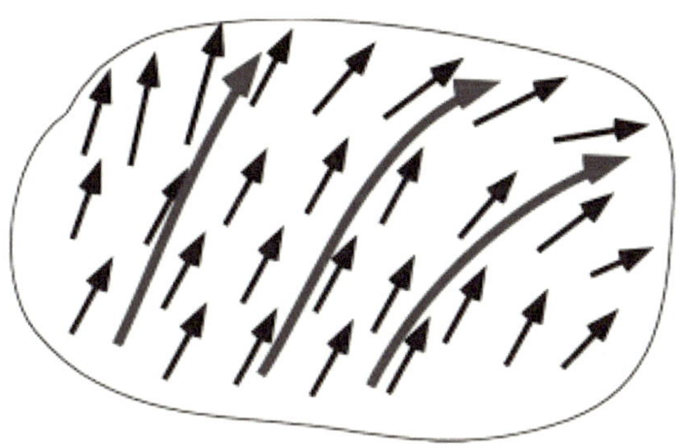

Figure 4.
Vector field and its flow.

$$[X, Y] = \left.\frac{d}{dt}\right|_{t=0} T\phi^X_{-t}\left(Y \circ \phi^X_t\right). \tag{14}$$

The Lie derivative is related to the derivation of functions by

$$[X, Y].f = X.(Y.f) - Y.(X.f) \tag{15}$$

where $X.f = Tf(X)$.

Now that we have introduced all necessary tools, we will talk about Riemannian geometry. The idea behind Riemannian geometry is to consider a scalar product that is dependent of the position, this will simplify the formalism, provided that we know how to manipulate the covariant differential calculus, which we will present right now.

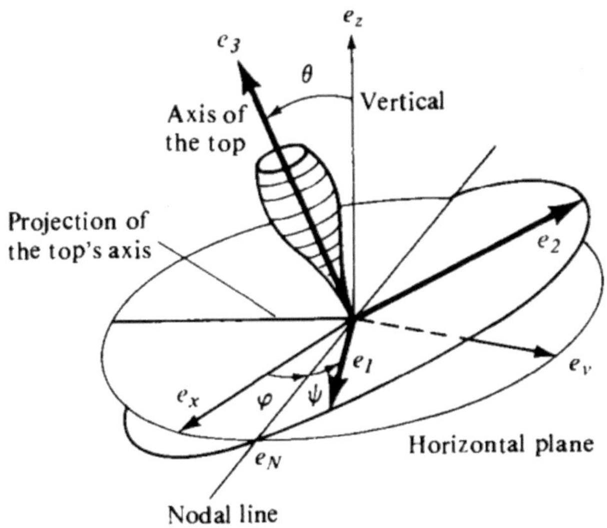

Figure 5.
The rigid body.

The kinetic energy of a ball that is constrained to stay in a surface S is independant of the position. In the general case, when we compute the kinetic energy of a rigid body or two link robot, this energy will depend of the configuration, this gives a Riemannian metric that is inherited from the inertia of the robot.

The innovative viewpoint of Arnold [6] is to see the motion of a rigid body fixed on one point as a curve in $SO(3)$, the kinetic energy depends only on the left angular velocity[7] and thus we say that the Riemannian metric is left invariant (**Figure 5**).

Return to our ball in a surface, in the absence of forces and potential, the principle of least action postulates that the real trajectory is a critical point of the kinetic energy, this means that the acceleration of the ball must be orthogonal to the surface, such trajectories are called geodesic, remark that if the ball was not constrained in S, this condition would give that the acceleration is 0, and thus the motion is a straightline, let us do the calculation

the kinetic energy for a ball with mass m is

$$E(\gamma) = \int_0^1 \frac{1}{2} m |\dot{\gamma}(t)|^2 dt \tag{16}$$

since $\gamma(t) \in S$, the authorized perturbations are $\gamma_s(t) \in S$ such that $\gamma_0(t) = \gamma(t)$, $\gamma_s(0) = \gamma(0)$ and $\gamma_s(1) = \gamma(1)$.

The postulate is that

$$\left.\frac{d}{ds}\right|_{s=0} E(\gamma_s) = 0 \tag{17}$$

this gives

[7] $\Omega = R^{-1}\dot{R}$.

$$\int_0^1 m\dot{V}(t).\dot{\gamma}(t)dt = 0 \tag{18}$$

where $V(t) = \frac{\partial}{\partial s}\big|_{s=0}\gamma_s(t) \in T_{\gamma(t)S}$ after integrating by parts we get

$$\int_0^1 V(t).\gamma''(t)dt = 0 \tag{19}$$

since $V(t) \in T_{\gamma(t)}S$, we get that

$$\gamma''(t) \perp T_{\gamma(t)}S. \tag{20}$$

The most important step in these calculations is the following one

$$\frac{d}{dt}X(t).Y(t) = \frac{dX}{dt}.Y(t) + X(t).\frac{dY}{dt} \tag{21}$$

where X, Y are tangent vector field to $\gamma(t) \in S$, that is $X(t), Y(t) \in T_{\gamma(t)}S$. In this case the kinetic energy which is the Euclidean scalar product does not depend on the position, for our applications, the kinetic energy depends on the position so we need to construct a "good derivative" that allow us to derive scalar products by computing the product of derivative only without computing the derivative of the metric.

The parallel transport [7] of a vector $v \in T_qM$ along a curve $\gamma(t) \in M$ is defined by the following way: the point at the origin of the vector follows the curve and the vector evolves continuously by preserving its length and the angle with $\dot{\gamma}(t)$.

This process defines a mapping $P(\gamma(1), \gamma(0)) : T_{\gamma(0)}M \to T_{\gamma(1)}M$ which is a linear isometry (preserving angle and length) (**Figure 6**).

The convariant derivative of two vector fields is

$$\nabla_X Y(q) = \frac{d}{dt}\bigg|_{t=0} P\left(q, \phi_t^X(q)\right) Y\left(\phi_t^X(q)\right) \tag{22}$$

this covariant derivative satisfies

$$\nabla g = 0 \tag{23}$$

which means that

$$X.g(Y, Z) = g(\nabla_X Y, Z) + g(Y, \nabla_X Z), \tag{24}$$

we say that this connexion kills the metric, in addition of that, this connexion is without torsion, that is we can close locally the parallelogram by doing parallel transport along geodesics by two different paths, the mathematical formula is

$$T(X, Y) = \nabla_X Y - \nabla_Y X - [X, Y] = 0 \tag{25}$$

for the computations in coordinates, we defines

$$\nabla_{\partial_i}\partial_j = \Gamma_{ij}^k \partial_k \tag{26}$$

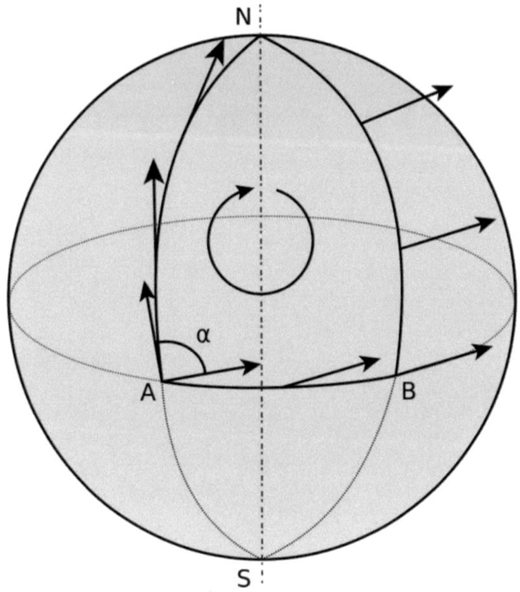

Figure 6.
Parallel transport.

the previous identities are[8]

$$\Gamma_{ij}^{k} = \Gamma_{ji}^{k}, \quad T = 0 \tag{27}$$

$$\Gamma_{ij}^{k} = \frac{1}{2} g^{km} \left(g_{mj,i} + g_{mi,j} - g_{ij,m} \right), \quad \nabla g = 0 \tag{28}$$

we construct a covariant derivative $\frac{D}{Dt} : \Gamma(\gamma) \to \Gamma(\gamma)$ along curves $\gamma(t) \in M$ that satisfies

- $\frac{DX+Y}{Dt} = \frac{DX}{Dt} + \frac{DY}{Dt}$
- $\frac{DfX}{Dt} = f'X + f\frac{DX}{Dt}$ for scalar function f
- $\nabla_{\gamma'(t)} V(\gamma(t)) = \frac{DV \circ \gamma(t)}{Dt}$ for $X \in \Gamma(TM)$.

In coordinates we get by setting $X(t) = X^i(t)\partial_{i|\gamma(t)}$

$$\frac{DX}{Dt} = \left(\frac{d}{dt}(X^i) + \Gamma_{kl}^{i} \frac{d}{dt}(q^k) X^l \right) \partial_{i|\gamma(t)}. \tag{29}$$

This covariant derivative satisfies

$$\frac{d}{dt} g(V, W) = g\left(\frac{DV}{Dt}, W \right) + g\left(V, \frac{DW}{Dt} \right). \tag{30}$$

[8] The component of the metric in local coordinates are $g_{ij}(q) = g_q(\partial_{i|q}, \partial_{j|q})$.

Now we suppose that our robot must evolves from q_1 to q_2 in M by minimizing the kinetic energy, we consider the real trajectory $q(t)$ and its admissible variations $q_s(t)$, by deriving we obtain

$$\frac{d}{ds}\Big|_{s=0}E(q_s) = \frac{1}{2}\int_0^1 g_{q_s(t)}(\dot{q}_s(t), \dot{q}_s(t))dt = \int_0^1 g_{q_s(t)}\left(\frac{D}{Ds}\Big|_{s=0}\dot{q}_s(t), \dot{q}(t)\right)dt \tag{31}$$

using the Schwarz lemma

$$\frac{D}{Ds}\frac{\partial}{\partial t} = \frac{D}{Dt}\frac{\partial}{\partial s} \tag{32}$$

we get

$$\int_0^1 g_{q(t)}\left(\frac{D}{Dt}V(t), \dot{q}(t)\right)dt = 0 \tag{33}$$

which gives by integration by parts that

$$\frac{D}{Dt}\dot{q} = 0 \tag{34}$$

which is the geodesic equation which are curves that are parallel along themselves.

Now we will talk about curvature, curvature is the opposite of flatness, in an Euclidean space, when we look at the parallel transport along closed curves, the obtained linear map is always identity, in a sphere, the set of obtained linear mappings cover all $SO(2)$, because of the curvature of the sphere.

So the curvature is the change of a vector by parallel transport along the parallelogram obtained by geodesics and parallel transport, the mathematical formula is

$$R(X, Y)Z = \nabla_X\nabla_Y Z - \nabla_Y\nabla_X Z - \nabla_{[X,Y]}Z \tag{35}$$

the curvature is a tensor, that is we can compute $R(u, v)w$ for $u, v, w \in T_qM$ without needing any extention.

The sectional curvature is

$$K(u, v) = \frac{g(R(u, v)v, u)}{|u|^2|v|^2 - g(u, v)}, \tag{36}$$

and it is the Gaussian curvature of the geodesic surface generated by (u, v).

Curvature is also a factor in the sensitivity with respect to initial conditions, the geodesic deviation is related to the sign of curvature by the following formula

$$\frac{D^2 J}{Dt^2} + R(J, \dot{q})\dot{q} = 0 \tag{37}$$

where J is a virtual motion that represents a local variation of geodesic (**Figure** 7).

When curvature is 0, the sensitivity is linear, which allow the Luenberger observer to be performant, when the curvature is negative, the geodesic instabillity is exponential [7], which explains that the Luenberger observer diverges (**Figure** 8).

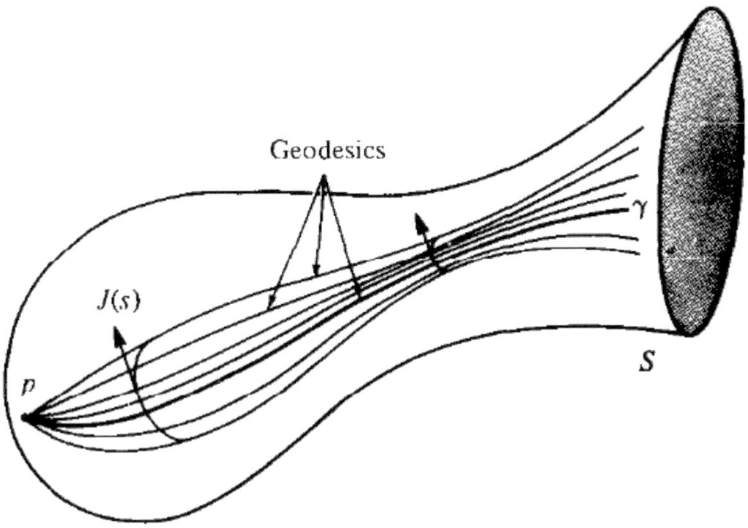

Figure 7.
Geodesic variation.

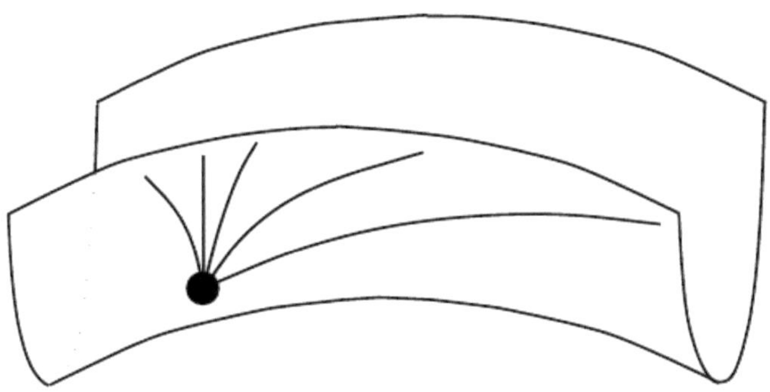

Figure 8.
Negative curvature and sensitivity of the geodesic flow.

The exponential map corresponds to a tangent vector the range by the geodesic after a unitary time, we denote it $\exp_q : V_q \subset T_qM \to M$, the exponential is a local diffeomorphism, so around a configuration, each configuration can be joined by a unique geodesic, the parallel transport along this geodesic is denoted by P (**Figure 9**).

when the exponential map is defined on T_qM, we say the (M,g) is complete, it is the fact when M is compact.

The geodesic distance is a natural distance induced by the Riemannian metric

$$d_g(q_1, q_2) = \inf_{\text{curves in } M \text{ joining } q_1 \text{ to } q_2.} \int_0^1 \sqrt{g_{q(t)}(\dot{q}(t), \dot{q}(t)} dt. \tag{38}$$

Thanks to Cauchy-Schwarz inequality, we can show that

$$\frac{1}{2} d_g(q_1, q_2)^2 = \frac{1}{2} \inf_{\text{curves in } M \text{ joining } q_1 \text{ to } q_2.} \int_0^1 g_{q(t)}(\dot{q}(t), \dot{q}(t) dt. \tag{39}$$

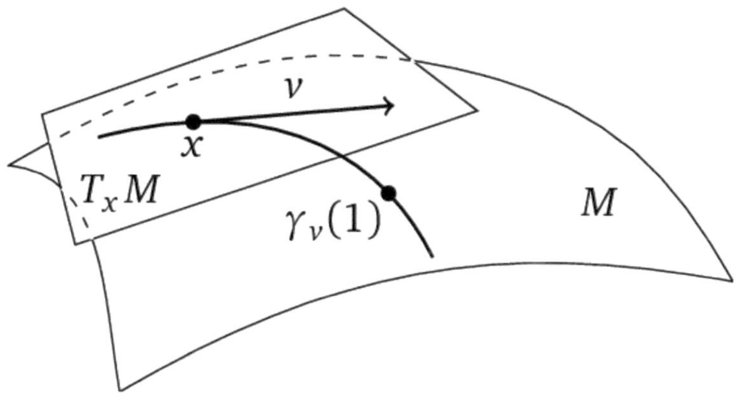

Figure 9.
Exponential map.

The gradient of a function f is the unique vector field ∇f that satisfies

$$df(X) = g(\nabla f, X) \tag{40}$$

The gradient of the function $U(q,.) = \frac{1}{2}d_g^2(q,.)$ is

$$\nabla U(p) = -\exp_p^{-1}(q). \tag{41}$$

The configuration space of the robot is a n dimensional compact manifold equipped with a Riemannian metric g that is the kinetic energy. In the presence of potential, the claculus of variations must applies to the Lagrangian $L(q,v) = \frac{1}{2}g(v,v) - W(q)$ and this gives the equation

$$\frac{D\dot{q}}{Dt} = -\nabla W(q) \tag{42}$$

in robotics, we have a control law that allow us to enslave the robot, so the equation becomes [8, 9]

$$\frac{D\dot{q}}{Dt} = -\nabla W(q) + u \tag{43}$$

We recover a Newton-like equation, the left term quantifies an obstruction of parallelism, and the right hand term an exterior force, as well as the Newton fundamental law "The trajectory of an object remains parallel along itself as long as there are no forces applying to it" (**Figure 10**).

The kinetic energy theorem gives the variation of the mechanical energy in function of the applied forces, this will be useful to compute feedback law to ensure regulation (**Figure 11**)

$$\frac{d}{dt}\left(\frac{1}{2}|\dot{q}(t)|^2_{q(t)} + W(q(t))\right) - g_{q(t)}(\dot{q}(t), u(t)). \tag{44}$$

In some applications, especially those concerning the motion of a rigid body, the configuration space is a Lie group G, which is a group furnished with a manifold structure

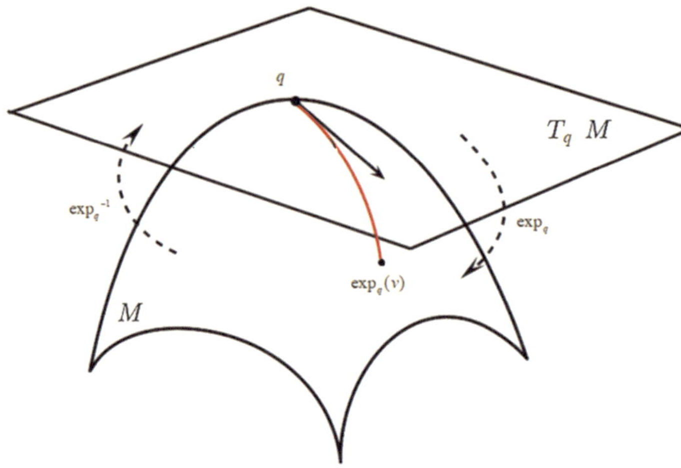

Figure 10.
Inverse of the exponential.

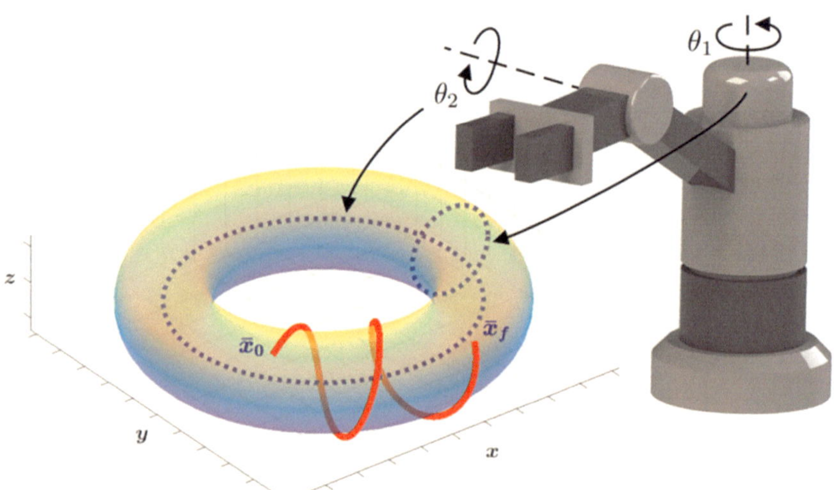

Figure 11.
Two link manipulator and its configuration space.

such that multiplication and inverse are smooth, the consideration of a Riemannian metric that is left invariant allow us simplificate the geodesic equation [9], the curvature tensor and the parallel transport, and thus simplificate the practical implementation of the optimal regulator [10, 11] and the Riemannian observer [12].

4. Regulation under constraints

In this section, we will propose a method to ensure regulation under constraints, before going further, lets see the case of regulation without constraints.

Let fix some reference configuration $q^* \in M$, and consider $W = 0$, by taking

$$u(t) = \exp^{-1}_{q(t)}(q^*) - k\dot{q}(t) \tag{45}$$

we get

$$\frac{d}{dt}\left(\frac{1}{2}|\dot{q}(t)|^2_{q(t)} + \frac{1}{2}d_g^2(q(t), q^*)\right) = -k|\dot{q}(t)|^2_{q(t)} \tag{46}$$

which ensure regulation $q(t) \to q^*$ by Lasalle's invariance principle, this result is very classical, but what we will prove in Section 6, is that this Riemannian PD-regulator is optimal for a natural cost.

Now we are interested by the regulation of the tool of a robot [13], the tool is the terminal organ of the robot, the control of the tool of a robot is frequent tackled problem in industry (**Figure 12**).

The position of the tool is modelised by a function $x : M \to \mathbb{R}^3$, we call the workspace $W_s = x(M)$ the range of x, that is the set of all possible positions that can be tooken by the tool.

A singular configuration is a configuration where the tool can not moves locally in all directions, that is a configuration $q \in M$ where $T_q x$ is not onto.

We have the following result that relates absence of singularities to the possibility of regulation of the tool [10].

Proposition 2 Suppose that the tool function is without singularities, take $x_d \in W_s$ and

$$u = -\nabla V - kv \tag{47}$$

where $V(q) = \frac{1}{2}\|x(q) - x_d\|^2$, *this controle stabilize the tool to* x_d *whith* 0 *velocity.*

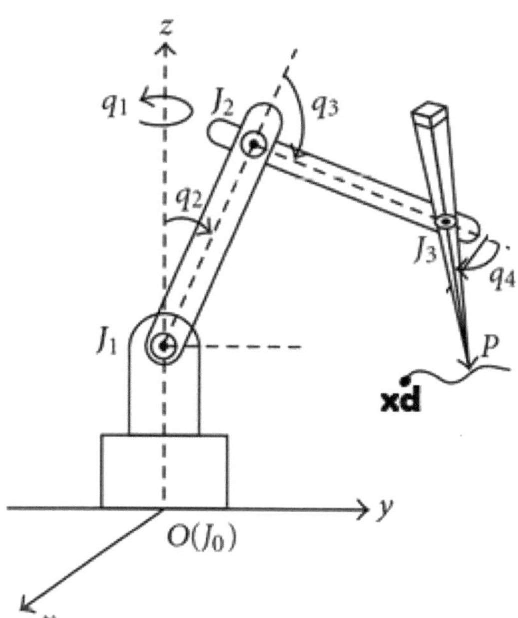

Figure 12.
Regulation of the tool.

Proof E: quilibrium configurations of this dynamics are critical points of V, which are $q \in M$ such that

$$T_q V(v) = (x(q) - x_d).dx_q(v) = 0 \tag{48}$$

since q is not singular, we must have $q \in x^{-1}(x_d)$.

Since the function $\frac{1}{2}|v|_q^2 + V(q)$ satisfies Lasalle's invariance principle hypethesis ans

$$\frac{d}{dt}\left(V(q(t)) + \frac{1}{2}|\dot{q}(t)|_{q(t)}^2\right) = -k|\dot{q}(t)|_{q(t)}^2 \leq 0 \tag{49}$$

The large subset invariant subset $M \times 0$ is $\Omega = x^{-1}(x_d) \times 0$. W

Now we turn to the regulation under constraints [13], if one wants to ensure regulation of the robot that leaves the tool in a safety area $S = \Phi^{-1}(0)$ where $\Phi : \mathbb{R}^3 \to \mathbb{R}$ is a submersion on $x(M)$ **(Figure 13)**.

We reduce the problem $x(q(t)) \in S$ to $q(t) \in N$ where $N = x^{-1}(S)$ is a hypersurface of M, the same calculus of variations done in the previous section allow us to see that

$$\frac{D\dot{q}}{Dt} + \nabla U(q) = C \in T_q N^{\perp} \tag{50}$$

C can be seen as a contact force which force the configuration to stay in N **(Figure 14)**

Thanks to an inverse calculation [10] one can show that the expression of the unique orthogonal component that allow the tool to remain in s is given by

$$u(q,v) = \frac{-<D_v \nabla\Psi(q), v>}{|\nabla\Psi(q)|^2} \nabla\Psi(q), \tag{51}$$

where $\Psi = x \circ \Phi$.

In some papers, the problem of tracking is tackled, and the most useful result is this one [14].

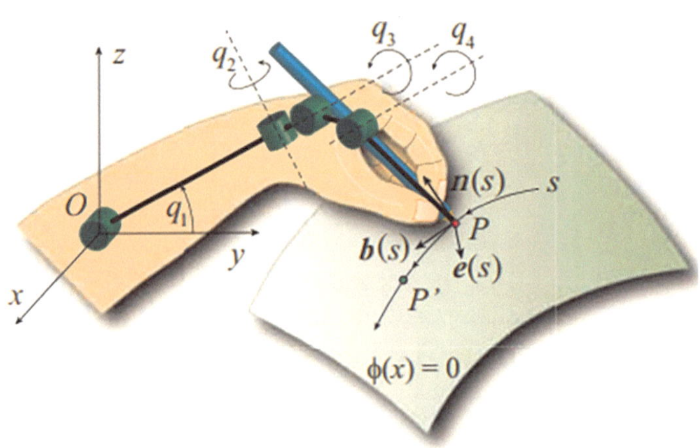

Figure 13.
Regulation under constraints.

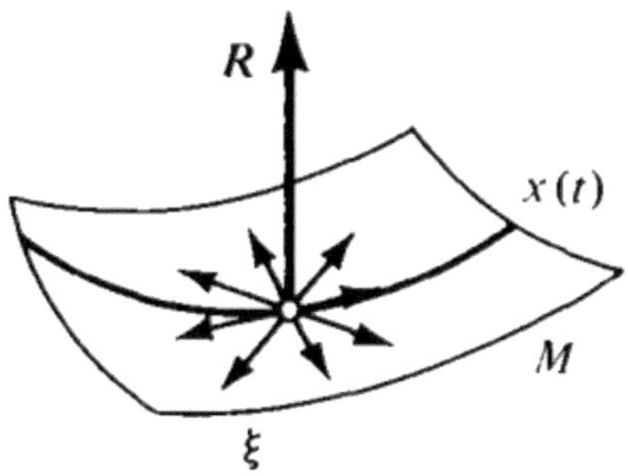

Figure 14.
Contact force.

Proposition 3 *For the dynamic* $\frac{Dq'}{Dt} = u$, *let a smooth trajectory reference on M, the regulator* $u = u_{FF} + u_{PD}$ *decreases the function*

$$t \in \mathbb{R} \to \frac{1}{2} d_g^2 \left(q_{ref}(t), q(t) \right) + \frac{1}{2} \left| v(t) - P(q(t), q_{ref}(t)) v_{ref}(t) \right|^2 \tag{52}$$

where

$$u_{FF}(q, v) = \frac{d}{dt} P\left(q, q_{ref}(t)\right) v_{ref}(t) + D_v P\left(q, q_{ref}\right) v_{ref}, \tag{53}$$

$$u_{PD} = \exp_q^{-1}\left(q_{ref}(t)\right) - k.\left(v - P\left(q, q_{ref}\right) v_{ref}\right). \tag{54}$$

5. The Riemannian observer

Another application of this formalism is the conception of observers of velocity. We want to get an estimation $t \to \hat{v}(t)$ of the velocity with the only data of the configuration $q(t)$.

We have explained in the second section that negative curvature is closely related to the sensitivity to the initial conditions of the geodesic flow. So for free systems, if we do not take into account the curvature, the observer will diverge if the observed initial velocity is far from the initial real velocity, even if the two configurations are very close.

The idea [15] is to copy the dynamic of the system and add a correction that cancel the curvature term that appears naturally in the derivation of the linearized equation around the real trajectory (Jacobi equation when absence of external forces). The observer of the dynamic

$$\frac{Dq'}{Dt} = S(q),$$

is

$$\frac{d\hat{q}}{dt} = \hat{v} + k_1 \exp_{\hat{q}}^{-1}(q(t)),$$

$$\frac{D\hat{v}}{Dt} = P(\hat{q}, q(t))(S(q(t))) + R\left(\hat{v}, \exp_{\hat{q}}^{-1}(q(t))\right)\hat{v} + k_2 \exp_{\hat{q}}^{-1}(q(t)).$$

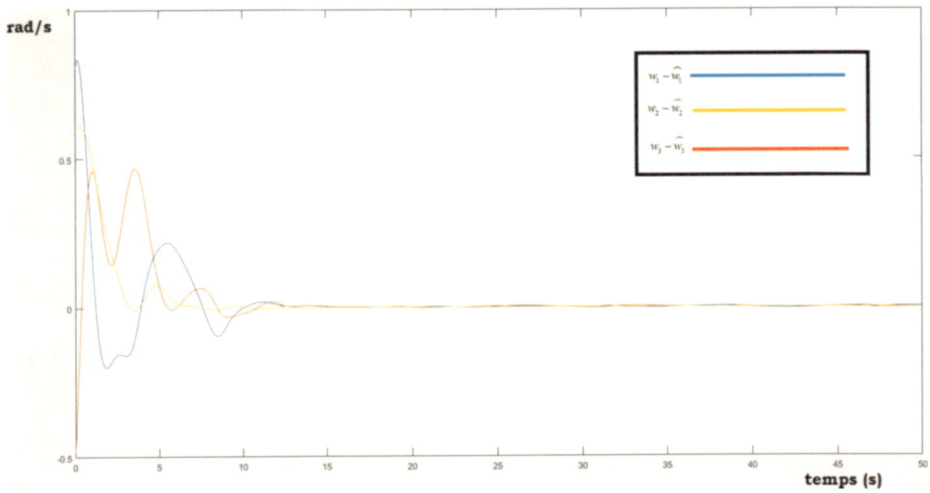

Figure 15.
Convergence of the Riemannian observer.

We can compute the form of this observer in the case of a Lie group furnished with a left invariant metric [12], this will give for a rigid body dynamic[9]

$$\dot{\hat{R}} = \hat{R}.(\hat{\Omega} - 2j(\zeta)), \tag{55}$$

$$\begin{aligned} \dot{\hat{\omega}} &= I^{-1}.[(I.\hat{\omega}) \times \hat{\omega} - (I\hat{\omega} \times \zeta + I\zeta \times \hat{\omega})] + \zeta \times \hat{\omega} - 2\zeta - 2.\hat{R}^T.R. \\ & I^{-1}[skew(R)]^{\times} + R(\hat{\omega}, \zeta)\hat{\omega}, \end{aligned} \tag{56}$$

with $\zeta = I^{-1}.\left[log\left(R^T.\hat{R}\right)\right]^{\times}$.

With the same conditions as the Luenberger observer of the first section, we get for the Riemannian observer the following simulations (**Figure 15**).

6. Optimal regulation and tracking

In this section we are concerned by the optimal control of robots, we consider the dynamics

$$\frac{D\dot{q}}{Dt} = u \tag{57}$$

and we consider the cost

$$J(u) = \int_0^{\infty} \left[\frac{1}{2}|v_u|^2\right]_g + \frac{1}{2}d_g(q_u(t), q^*)^2 + |u|_g^2]e^{-\gamma t}dt \tag{58}$$

where (q_u, v_u) is the unique solution to the dynamics (after fixing an initial position and velocity).

---

[9] $\log(R) = \frac{\phi(R)}{\sin(\phi(R))} skew(R)$, with $\phi(R) = \arccos\left(\frac{tr(R)-1}{2}\right)$.

The HJB theory gives that the feedback control [10]

$$u_{PD} = k \exp^{-1}_{q(t)}(q^*) - k'\dot{q}(t) \tag{59}$$

where k, k' solves an appropriate Riccati equation, is the unique that solves the previous problem.

For the tracking problem, we have the following result [10].

Proposition 4 *Let a smooth reference trajectory* $q_{ref} : [0, T] \to M$, *the tracking regulator that minimizes the cost*

$$J(u) = \int_0^T \left[\frac{1}{2} d_g^2\left(q_{ref}(t), q_u(t)\right) + \frac{1}{2}\left|v_u(t) - P(q_u(t), q_{ref}(t))v_{ref}(t)\right|^2 + \frac{1}{2}|u - u_{FF}(t, q, v)|^2\right] dt, \tag{60}$$

is a PD + Feed forward regulator $u = u_{FF} + u_{PD}$ where

$$u_{FF}(t, q, v) = \frac{d}{dt} P\left(q, q_{ref}(t)\right)v_{ref}(t) + D_v P\left(q, q_{ref}(t)\right)v_{ref})(t), \tag{61}$$

$$u_{PD}(t, q, v) = k_1(t) \exp^{-1}_q\left(q_{ref}(t)\right) - k_2(t)\left[v - P\left(q, q_{ref}\right)v_{ref}(t)\right], \tag{62}$$

avec (k_1, k_2) *solves an appropriate Riccati equation.*

For the applications to a rigid body motion [11, 16]

$$R' = R\Omega, \tag{63}$$

$$\omega' = I^{-1}.(I\omega \times \omega) + \tau. \tag{64}$$

We have the following PD that ensures optimal regulation to R_d

$$\tau_{PD} = -k_P\left[\log(R_d^T.R)\right]^\times - k_D\omega. \tag{65}$$

For the optimal tracking problem, the PD + FF regulator is

$$\tau_{PD} = -k_P\left[\log\left(R_{ref}^T R\right)\right]^\times - k_D.(\omega - R^T R_{ref}.\omega_{ref}), \tag{66}$$

$$\tau_{FF} = \frac{1}{2}(\omega \times (R^T R_{ref}.\omega_{ref}) - I^{-1}.((I.R^T R_{ref}\omega_{ref}) \times \omega + (I.\omega) \times (R^T R_{ref}\omega_{ref}))) \tag{67}$$

7. Conclusion

Classical robots allow us to ensure local regulation of the configuration and optimal control of kinematics of rigid body. However, some observations shows that it do not fit well with regulation under constraints and optimal regulation of the dynamics. We cannot explain the origin of the sensitivity with respect to initial conditions, which is present in most robotic systems and prevent us to design performant observers.

We have introduced the geometric formalism with a comprehensive way, and we explained the origin of the sensitivity with respect to initial conditions by the notion of curvature, that we avoid by the Riemannian observer.

We have exposed how to ensure regulation under constraints under a hypothesis of absence of singularities, and we showed how to ensure optimal regulation and tracking by algorithmic methods.

However, this method requires fastidious computations and sometimes difficult, approximations are quiete often used, and by doing that we see that we recover regulators that are proposed by the Euclidean approach.

We can see the classical robotics as an approximation of the geometric robotics, a bit like newtonian mechanics is the approximation of general relativity.

Geometric robotics is a field active in research, and most of recent results are presented in this chapter. Open problems remains in soft robotics, where we control a quasilinear hyperbolic PDE written in the Lie algebra of the group of Euclidean isometries, for this purpose one can see the work of [17] and the reference within.

Author details

Anis Bousclet
Université de Technologies de Troyes, Rosières-Près-Troyes, France

*Address all correspondence to: anis.bousclet@utt.fr

References

[1] Khalil W, Dombre E. Modélisation, identification et commande des robots. Paris: Hermes Science Publication; 1999

[2] Slotine J-J. Applied Nonlinear Control. Pearson; 1990

[3] Aguiar AP, Sacoon A, Hauser J. Exploration of kinematic optimal controlon the Lie Group SO(3). In: 8th IFAC Symposium on Nonlinear Control Systems. Vol. 43. Issue 14. 2010. p. 1302-1307

[4] Luenberger DG. Observing the state of a linear system. IEEE Transaction on Military Electronics. Vol. 8. New York. 1964. pp. 74-80

[5] Lee J. Introduction to Smooth Manifolds. Graduate Texts in Mathematics. 2012

[6] Arnold V. Mathematical Methods of Classical Mechanics. Graduate Texts in Mathematics. Vol. 60. New York: Springer; 1978

[7] Lee J. Riemannian Geometry: An Introduction to Curvature. New York: Springer-Verlag; 1997

[8] Altafini C. Geometric control methods for nonlinear systems and robotic applications [Ph.D thesis]. Sweden: Royal Institute of Technology, University of Stockholm; 2001

[9] Lewis AD, Bullo F. Geometric Control of Mechanical Systems. Texts in Applied Mathematics. 2005

[10] Bousclet A. Régulateurs Optimaux en Robotique. Mémoires de fin d'études de l'École Nationale Polytechnique; 2021

[11] Guo J, Liu C, Tang S. Intrinsic optimal control for mechanical systems on lie groups. Advances in Mathematical Physics. 2017:1-8

[12] Jordan M, Berg D, Maithripala HS. An Intrisic Observer for a Class of Simple Mechanical Systems on a Lie Group. Vol. 2. New Orleans USA: Proceedings of the American Control Conference; 2004. pp. 1546-1551

[13] Sekimoto M, Tahara K, Arimoto S, Yoshida M. A Riemannian geomtry approach for control of robotic systems under constraints. SICE Journal of Control, Measurement, and System Integration. 2021:107-116

[14] Murray RM, Bullo F. Tracking for fully actuated mechanical systems: A geometric framework. Automatica. 1999; **35**(1):17-34

[15] Aghannan N, Rouchon P. An intrinsic observer for a class of Lagrangian systems. IEEE Transactions on Automation and Control. 2003;**48**(6): 936-945

[16] Tayebi A, Berkane S. Some optimisation aspects on the lie group SO (3). IFAC-PapersOnLine. 2015;**48**(3): 1117-1121

[17] Bousclet A. Geometric Formulation of Dynamics of Snake Robot. Master thesis of Paris Saclay University; 2022

Section 4

Machine Learning and Vision

Chapter 6

Challenges and Trends of Machine Learning in the Myoelectric Control System for Upper Limb Exoskeletons and Exosuits

Jirui Fu, Zubadiah Al-Mashhadani, Keith Currier, Al-Muthanna Al-Ani and Joon-Hyuk Park

Abstract

Myoelectric control systems as the emerging control strategies for upper limb wearable robots have shown their efficacy and applicability to effectively provide motion assistance and/or restore motor functions in people with impairment or disabilities, as well as augment physical performance in able-bodied individuals. In myoelectric control, electromyographic (EMG) signals from muscles are utilized, improving adaptability and human-robot interactions during various motion tasks. Machine learning has been widely applied in myoelectric control systems due to its advantages in detecting and classifying various human motions and motion intentions. This chapter illustrates the challenges and trends in recent machine learning algorithms implemented on myoelectric control systems designed for upper limb wearable robots, and highlights the key focus areas for future research directions. Different modalities of recent machine learning-based myoelectric control systems are described in detail, and their advantages and disadvantages are summarized. Furthermore, key design aspects and the type of experiments conducted to validate the efficacy of the proposed myoelectric controllers are explained. Finally, the challenges and limitations of current myoelectric control systems using machine learning algorithms are analyzed, from which future research directions are suggested.

Keywords: myoelectric control, upper limb exoskeleton, upper limb exosuit, pattern recognition, machine learning, reinforcement learning

1. Introduction

In the past few decades, the demand for upper limb exoskeletons and exosuits has grown substantially due to their promising applications across industry, medical and military sectors. The exoskeletons consists of rigid links and joints attached to the human body, whereas the exosuits use soft and flexible materials (such as fabric or soft polymer) to interact with the user's body [1]. The applications of exoskeletons and exosuits include: (i) power augmentation to enhance physical performance or the capabilities of able-boded individuals during strenuous physical tasks [2], and (ii) assisting individuals with disabilities in performing activities of daily living (ADLs) [3].

The exoskeletons and exosuits can be controlled by many different schemes, such as the kinematics control based on the inertia measurement unit sensor (IMU) or encoder [4, 5], the force control based on load cell or torque sensor [6, 7], and the myoelectric control based on the electromyographic sensor (EMG) [8]. Among these control schemes, the myoelectric control systems have gained increasing attention over recent years [9–11]. The myoelectric control systems of the upper limb exoskeletons and exosuits use surface electromyography (EMG) signals, the electric potentials directly measured from the skeletal muscle as input of the control system for exoskeletons and exosuits (**Figure 1**). The surface EMG signals are generated from the motor unit activation, controlled by the human brain, and regulated by the motor neurons in the spinal cord. The mechanism for generating surface EMG signals offers surface EMG signals to detect human movement intention [12]. The critical advantage of a myoelectric control system over other control systems is its timely detection of the user's motion intention leveraging electromechanical delay (EMD); the onset of motion can be detected about 50–100 ms earlier than the physical motion [13, 14]. Moreover, the exoskeletons and exosuits equip with myoelectric control systems have a more adaptive and intelligent interface with the users as the exoskeleton and exosuits can timely and proactively engage assistance through detecting the users' movement intention [15].

Myoelectric control systems for upper limb exoskeletons and exosuits initially used on-off/finite state control and proportional control, as described in Refs. [16, 17]. Although these methods are simple and easy to implement, their ability to accommodate a wide range of different movements is limited, as noted in Ref. [18]. Consequently, their primary use have been limited to a single joint function such as elbow flexion/extension or hand grip. To allow for more complex movements across multiple degrees of freedom (DOFs), machine learning (ML) and deep learning (DL) algorithms have been utilized in the myoelectric control systems. However, the myoelectric control systems with ML or DL algorithms generally require considerable computational power, which imposes practical limitations on the portability of exoskeletons and exosuits [19]. In recent years, with the advancements in more powerful and compact embedded computers, myoelectric control systems with ML or DL algorithms became feasible to implement on upper limb exoskeletons and exosuits. Compared to the early staged myoelectric control modalities, the ML or DL-based myoelectric control systems have shown superior performance and better results in complex, multi-DOF upper limb motions; yet, there still exist challenges and limitations which will be discussed in detail in the subsequent sections.

Given the growing interest in machine learning and deep learning-based myoelectric control systems for upper limb exoskeletons and exosuits, the number of publications in the relevant field has rapidly increased over the past decade. Therefore, it is imperative to understand the latest trend and challenges in machine learning and deep learning-based myoelectric control system for upper limb exoskeletons and exosutis. A systematic review that provides a comprehensive overview of the myoelectric control system for upper limb exoskeletons and exosuits [8] was published by the authors. However, the focus of that review was not specifically machine learning and deep learning-based myoelectric control system of upper limb exoskeletons and exosuits, and it does not discuss current challenges and future directions. This chapter

DOI: http://dx.doi.org/10.5772/intechopen.111901

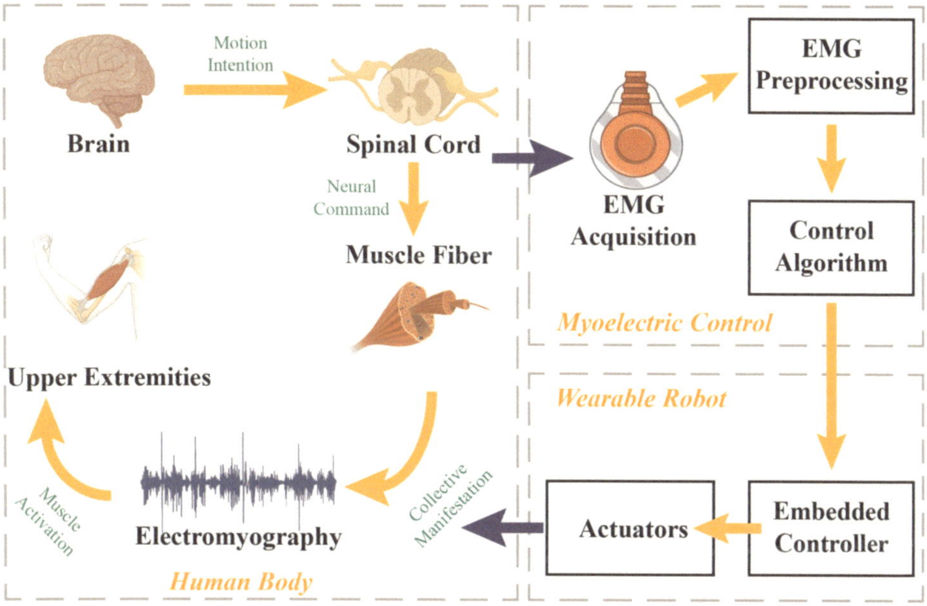

Figure 1.
The generation of electromyography signal and the workflow of myoelectric control systems on upper limb exoskeletons and exosuits.

is designed to share the extensive review of the machine learning-based myoelectric control system for upper limb exoskeletons and exosuits, particularly from scientific articles published between 2011 and 2023. The identified challenges in implementing machine learning algorithms in the myoelectric control system and future directions are suggested. In the following section, the process of machine learning-based myoelectric control system is summarized (Section 2), and the state-of-the-art implementation of machine learning algorithms in upper limb exoskeletons and exosuits is presented (Section 3). Finally, the remaining unaddressed research questions and tasks are discussed as future research directions (Section 4).

2. The procedure of machine learning-based myoelectric control system

Similar to any other types of myoelectric control systems, machine learning-based myoelectric control systems include key procedural steps: signal acquisition, pre-processing, feature extraction, and motion intention detection through the trained machine learning model (**Figure 2**). In this section, the process of a machine learning-based myoelectric control system will be presented in detail.

2.1 Data acquisition and signal processing

The acquisition of the EMG signal is critical to the myoelectric control system of the upper limb exoskeleton, as the accuracy of the myoelectric controller primarily depends on the quality of the EMG signal. It is, therefore, important to acquire quality and accurate EMG signals. Three essential components of EMG data acquisition systems are the electrodes for EMG, the sampling rate and signal filtering.

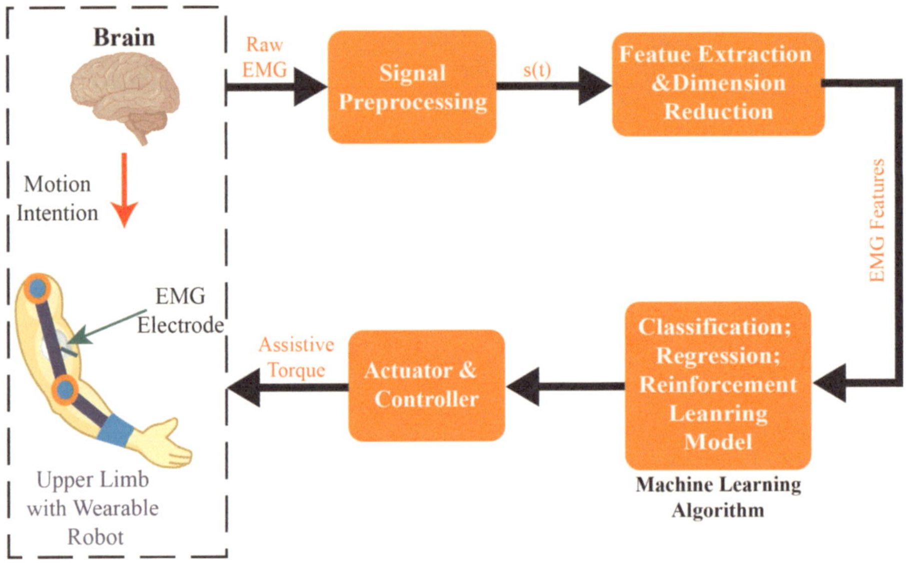

Figure 2.
The process of machine learning based myoelectric control system.

- EMG electrodes – The electrodes of the EMG sensors include surface and intramuscular, in which the surface EMG electrode uses an insulative sticker to place the electrode on the skin overlying a muscle to detect the electrical activity of the muscle [20]. On the other hand, the intramuscular EMG electrodes utilize the needles or wires that are inserted into the muscles [21]. Compared to the surface electrode, the intramuscular EMG electrodes can minimize the crosstalk from neighboring muscle segments; however, the operation of intramuscular EMG electrodes requires trained medical expertise, and it is more difficult to use in the research environment. Thus, the surface EMGs are more commonly and widely used in myoelectric control systems of upper limb exoskeletons and exosuits.

- Sampling rate – Because the acquisition of EMG signal involves converting the analog signal (voltage generated by muscular activation) to the digital signal (EMG signal used in the myoelectric control systems), which a computer can process, selecting an optimal sampling rate is critical to avoid under-sampling or oversampling. The choice of sampling rate varies between muscle segments. For example, [22] investigates the selection of sampling rate for EMG, and [23] explores the effect of sampling rate for machine learning-based myoelectric control system accuracy

- Filtering and rectification – The acquisition of surface EMG measures a combination of the activation of all recruited motor units within the muscle. Therefore, the collected surface EMG signal contains the drift and artifacts which affect the accuracy of the surface EMG signal. The Butterworth filters have been widely used to remove the drift and artifact from collected surface EMG signals. Usually, filtering the raw surface EMG signal includes using a high-pass Butterworth filter to remove the drift and artifact from the raw surface EMG signal, then using a low-pass Butterworth filter to acquire an envelope indicating the magnitude of the surface EMG signal as it changes over time. However, the selection of the order and cut-off frequency of the Butterworth filter could be optimized for different muscle segments; for example, [24] presents the filter selection for surface EMG signal to remove the drift and artifacts. After high and low pass (or band pass) filters, the EMG signals need full wave rectification then a low pass filter for further processing and feature extraction.

2.2 Feature extraction

The pre-processed surface EMG signal is presented as a time sequence that includes a large number of randomness. Therefore, directly feeding the pre-processed surface EMG signal to the machine learning model is impractical. To feed the pre-processed surface EMG signal to the machine learning model, the sequence of pre-processed surface EMG signals must be mapped into a smaller dimension vector called a feature vector [25]. The process of extracting feature vectors from the pre-processed surface EMG signal is called feature extraction. In applying a myoelectric control system for upper limb exoskeletons and exosuits, feature extraction includes two types of methods: feature selection and dimensionality reduction algorithms.

- Feature selection – The feature selection uses the statistic formulas to convert the pre-processed surface EMG signal to low-dimensional feature vectors. The feature selection methods are categorized as time-domain and frequency-domain features. However, according to the literature review, only the time-domain features are utilized in the myoelectric control systems for upper limb exoskeletons and exosuits because the time-domain features are computationally simple compared to the frequency-domain features. The time-domain features use the time sequence of surface EMG signal, and there are many types of time-domain features used such as integrated EMG (IEMG), mean absolute value (MAV), root mean square (RMS) and auto-regressive coefficient (AR). Among these, the root mean square and auto-regressive coefficient features are the most widely used time-domain EMG features in myoelectric control systems for upper limb exoskeletons and exosuits, according to our literature survey. Because each feature has different statistical meanings and implications, the effect of selecting different features for the machine learning model could vary. For example, [26] investigated the effect of different features on hand motion classification.

- Dimensionality reduction – Similar to the statistical features, the dimensionality reduction also maps the pre-processed surface EMG signal to the low-dimensional feature vectors. However, the statistical formulas used in feature selection do not change the data, while the dimensionality reduction maps the data to the lower dimension. The dimensionality reduction uses linear or nonlinear algorithms to map the data to a low dimension, such as the Principle Component Analysis (PCA) and Linear Discriminate Analysis (LDA). In [27], the performance of dimensionality reduction in classifying the object's weight using a machine learning model from the surface EMG signal was studied. The dimensionality reduction algorithms are less frequently used in myoelectric control systems for upper limb exoskeletons and exosuits due to their comparably heavy computational process as compared to feature selection process.

3. Taxonomy of machine learning-based myoelectric control systems

The previous section presents an overview of the process of machine learning-based myoelectric control system. This section will summarize the machine learning algorithms used in the reviewed research articles. These algorithms are categorized into three: (1) classification-based models, which are used to detect the type of movement from the surface EMG features, (2) regression-based models, which can make a continuous prediction of the human subject's joint kinematics or torques, and (3) the reinforcement learning models which optimize the model through the interaction of human subjects and machine learning model. As shown in **Figure 3**, among the included research articles, the regression-based models and classification models are the most widely used modalities in the machine learning-based myoelectric control systems, while not much work has been done in the implementation and validation of reinforcement learning models.

3.1 Classification-based myoelectric control system

The classification-based Myoelectric Control System uses the classification model to detect the movement from the statistical features of the human subject's surface EMG signal. In the classification-based myoelectric control, the labels are pre-defined by the human subjects, which includes types of upper limb movement such as elbow flexion/extension (diversified labels), and the onset of upper limb movement such as in motion or still (binary labels). To train the classification models, surface EMG data corresponding to the labeled motion must be collected from human subject. Then, the classification model can be trained by various machine learning algorithms such as the Support Vector Machine (SVM) [28], Linear Discriminant Analysis (LDA) [29], K-nearest neighbors (KNN) [30], etc. According to our literature review, the support vector machine algorithm is the most popular choice in the classification models of machine learning-based myoelectric control systems. Compared to other machine learning algorithms, the support vector machine algorithm provides better computational efficiency that makes it feasible to run on embedded computers than other types of machine learning algorithms. The support vector machine algorithm can train classification models with either diversified or binary labels. For example, [31] trained

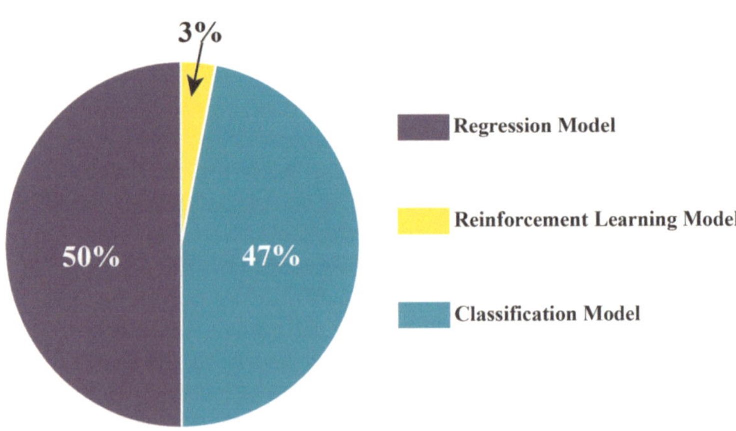

Figure 3.
Different types of machine learning models used in the reviewed research articles.

a SVM model to classify if the human subject's finger is in motion or not. In [31], a classification-based myoelectric control system trained by the SVM algorithm was proposed for a (#DOF) finger exoskeleton using binary labels. Additionally, [32] compared the accuracy of the classification-based myoelectric control systems for a hand exoskeleton trained by different machine learning algorithms, i.e., SVM, artificial neural network with backpropagation algorithm (ANN), and K-nearest neighbors (KNN). The classification model in [32] includes five labels that correspond to five different types of hand motion, then used the classification output to trigger the predefined assistive mode in the hand exoskeleton. According to [32], the classification model trained by SVM showed the best accuracy among those compared. Moreover, Cheon et al. proposed a myoelectric interface based on the musculotendinous junctions (MTJs) of the flexor digitorum superficialis (FDS) for reliable control of a robotic glove with a single EMG sensor by identifying power grasp intentions [33] and the support vector machine (SVM) algorithm was used to optimize the classification model. Other machine learning algorithms have also been utilized to train the classification model in the machine learning-based myoelectric control systems. For example, [34] utilized the MCLPBoost – a type of decision-tree algorithm to classify the flexion and extension of elbow and wrist joints. Compared to the SVM algorithm, they showed that the MCLPBoost had better robustness against the noised training data.

Many research articles reviewed targeted to improve the performance of classification models. For instance, [16, 35] studied the impact of feature extraction on the accuracy of classification model where two types of feature extraction techniques were explored. The type 1 feature extraction technique converted the single-channel EMG signal to 14 different statistical features; the type 2 feature extraction technique converted five channel EMG signal to a single statistical feature. Both type 1 and 2 feature extraction techniques were designed for the same upper limb exoskeleton and the classification models were trained by the same machine learning algorithm. The experimental result indicated the type 1 feature extraction technique outperformed the type 2 feature extraction from which they suggested that when training the classification-based myoelectric control systems, higher dimensional training set gives better performance. Moreover, [36, 37] implemented the sensor fusion method by combining the EMG and electroencephalography (EEG) signal to improve the accuracy of the classification-based myoelectric control system trained by the artificial neural network with a backpropagation algorithm. Additionally, to prevent the misclassification caused by the unfiltered noise in EMG signals such as crosstalk and motion artifacts, [38] utilized a threshold method in which the amplitude of filtered EMG signal must be greater than a specific value to be an input to the classification model. Twardowski et al. used the machine learning algorithm to convert the motor unit firings from the sEMG signals into biomechanically informed signals that drive the actuation [39]. The resulting signal provides a smoother control scheme with less delay versus using the MAV and RMS response to modulate the actuation. The EMG signal in the study [31, 32, 34, 39] used the root mean square (RMS) as statistical features, while [40] used integrated EMG (iEMG) to train the classification model. Compared to the RMS feature, the iEMG feature requires less computational power. The classification model presented in [40] plotted the output data onto a 2D Cartesian plane that can be distinguished in real-time using a Point-in Polygon algorithm commonly used in computer graphics. This algorithm determines whether the sample in the plane belongs in or out of a given polygonal area which is the area of each given label. Among the tested classifiers, this method provided the highest classification accuracy (94%) when classifying hand grasp motions.

The abovementioned articles utilized the statistical features of EMG data as input for the classification-based myoelectric control systems. However, the raw EMG signal can also be used as input for the classification-based myoelectric control system, as demonstrated by [41], which successfully implemented a vision transformer model to classify two datasets using raw multichannel EMG data. The transformer model is commonly used in natural language processing, but the encoder-decoder network can be applied to determine the underlying characteristics of the input data without manual feature extraction or signal pre-processing. The resulting model achieved a higher classification accuracy versus a convolution neural network model and an LSTM network.

3.2 Regression-based myoelectric control system

The regression-based myoelectric control system implements regression analysis techniques. In statistics, regression analysis estimates the relationship between a dependent variable (output of regression-based myoelectric control system) and an independent variable (usually the EMG features in the regression-based myoelectric control system) by using a regression model. Compared to the classification-based myoelectric control system, the regression model can output continuous variables such as joint torque and joint angle. The regression model can be trained by various machine learning algorithms. However, there are two regression models found in our literature review, artificial neural network with backpropagation algorithm and Kalman Filters.

Among the research articles reviewed, the artificial neural network was the most widely used method to train the regression model. For example, [42] implemented a regression model to estimate the joint angle from the statistical feature of the human subject's EMG signal. The regression model is trained by artificial neural network with a back propagation algorithm, and the results showed that the regression model could accurately estimate the joint angle of human. Additionally, the regression-based myoelectric control systems have also been widely used in the bilateral training of hand exoskeletons. Because the bilateral training focuses on using the unimpaired hand to help the impaired hand restore its motor control capability, the myoelectric control scheme must accurately estimate the joint kinematics or joint torque of the unimpaired hand which complies with the characteristics of regression-based myoelectric control systems. For example, [38, 43–48] implemented the regression model to estimate the joint angle or joint torque from the unimpaired hand to help the impaired hand to restore its motor control capability. On the other hand, Kalman Filter is another approach used in the regression model for myoelectric control of upper limb wearable robots. Compared to the artificial neural network with backpropagation method, Kalman filter does not need much time and extensive datasets to train the model. Moreover, tuning Kalman filter requires less computational power than tuning the artificial neural network which makes it easier to run on an embedded computer. The studies [49, 50] utilized the Kalman filter to compute the joint torque based on the EMG signal whose regression models offered better accuracy when compared to the regression model trained by artificial neural network with backpropagation algorithm. Another method proposed by Kopke et al. used 6 DOF loadcells and EMG sensors to acquire the training data and the linear discriminate analysis (LDA) algorithm to train the regression model [51]. The experiment demonstrated a 92% accuracy in estimating the joint torque of human subjects' shoulder and elbow joint.

Furthermore, some studies focused on improving the accuracy of regression models. For example, Sierotowice et al. [52] utilized a ridge regression algorithm and a feature selection algorithm called Random Fourier Features to improve the accuracy of the regression model to estimate the hand-grasping force. The regression algorithm of the controller achieved a higher classification accuracy when determining the target forces versus the random Fourier features algorithm (80% versus 73%, respectively). Moreover, the work by Meattini et al. used a soft dynamic time warping (soft-DTW) method to improve the accuracy of the neural network based regression model [53] and the result of this study shows comparable performance to the conventional neural network regression model.

3.3 Reinforcement learning based myoelectric control system

The reinforcement learning algorithm is another type of machine learning algorithm which are used as a machine learning based myoelectric control system. Different from the classification and regression models, the reinforcement learning model trains an agent to choose the optimal action under a specific state in an environment. The process of reinforcement learning can be divided into several steps; in each step, the smart agent executes an action based on a specific state and receive a reward signal as feedback. The objective of the smart agent is to find the optimal action to maximize the accumulative reward.

Compared to the other two types of machine learning myoelectric control systems, only a few included research literature implemented the reinforcement learning algorithm. Hamaya et al. [54] utilized an elbow exoskeleton and applied the Probabilistic Inference for Learning Control (PILCO) reinforcement learning algorithm. The state vector included elbow joint kinematics and EMG signals, and the reward was based on the deviation between the intended and actual trajectory. PILCO employed the Gaussian process to learn the probabilistic dynamic model of the interface between the human and the exoskeleton. The learned model was then used to assess the control policy, which was optimized using the policy gradient method [55]. This approach proved to be more efficient than other machine learning myoelectric control systems, leading to a shorter training period.

4. Discussion

This section outlines several research questions and tasks that need to be addressed in future studies, including the robustness of machine learning-based myoelectric control system, the incorporation of safety requirements in machine learning-based myoelectric control systems, and the clinical assessment of assistive and rehabilitative upper limb exoskeletons and exosuits with machine learning-based myoelectric control systems. These research questions point out crucial barriers to the effective use of machine learning-based myoelectric control systems in upper limb exoskeletons and exosuits which warrant further investigations.

4.1 Robustness of machine learning-based myoelectric control systems

The myoelectric control system's ability to withstand disturbance from both internal and external sources within the environment, as measured by its resistance to electromyography signals [56], is referred to as its robustness. This type of disturbance is typically caused by muscle fatigue [57], electrode displacement [58], and changes in EMG patterns over time [59]. Over the past decade, there has been a significant increase in studies employing machine learning-based myoelectric control systems, which have shown promising results in preliminary or pilot testing in laboratory settings. However, none of these systems have explored methods to enhance their robustness. To bridge the gap between experimental research and commercial or clinical applications, machine learning-based myoelectric control systems should concentrate on creating a precise control scheme under well-controlled laboratory conditions while also improving robustness in real-world scenarios.

The review of research articles that utilized machine learning-based myoelectric control systems found that these systems face common issues, such as varying characteristics of sEMG signals in different physiological conditions, noise/artifacts, muscle fatigue that causes variance in sEMG signals, and electrode shift during or between sessions. However, none of the studies focused mainly on addressing these issues. Existing studies have investigated these issues in the context of myoelectric control of prosthetics, teleoperate robotic arms, and pattern recognition of sEMG signals. Potential approaches to improve the robustness of machine learning-based myoelectric control systems include using more efficient features, reducing the impact of EMG electrode shift, and improving the data collection protocol or signal processing method. However, these methods have not been studied in the included research articles. Therefore, further investigations are needed to evaluate the performance of machine learning-based myoelectric control systems with these robustness-improving methods and their performance on the upper limb exoskeleton.

In future studies, it is suggested to investigate the performance of upper limb exoskeletons with machine learning-based myoelectric control systems using different time-domain and frequency-domain features. The selection of EMG features should be expanded to account for larger time-domain and frequency-domain features, and the performance of the human-exoskeleton system with the improved myoelectric control system should be evaluated. Additionally, during laboratory research, the causes of error, such as EMG electrode shift and muscle fatigue that could affect the robustness of machine learning-based myoelectric control systems in clinical applications, should be emulated. Novel training protocols should also be investigated because using the EMG signal collected within a short period to train the machine learning-based myoelectric control system will affect its robustness. Therefore, future studies of machine learning myoelectric control systems of upper limb exoskeletons should focus on developing novel control schemes, investigating effective training protocols, and evaluating them on the upper limb exoskeletons. In the research articles reviewed, there were several common issues that were reported. These issues included differences in the characteristics of EMG signals across various physiological conditions, the presence of noise and artifacts, muscle fatigue leading to variations in EMG signals, and electrode movement during or between sessions. However, none of the included research articles specifically addressed these issues by focusing on improving the robustness of machine learning-based myoelectric control systems.

Furthermore, one of our studies explored the implementation of a variational autoencoder to improve the robustness of the classification model in using the EMG signal to recognize the motion performed by the human subject. An autoencoder is a neural network model that is trained to compress and uncompress inputted data while reducing the error between the input data and the reconstructed output data as much as possible [60]. The restrictive architecture of the autoencoder creates a model that can act as a dimensionality reduction method to perform unsupervised feature learning. Implementing autoencoder networks or more advanced encoder-decoder networks can further reduce the complexity of input myoelectric signal data or multimodal sensor data at the compressed latent layer while learning the hidden characteristics that define the system. Autoencoder networks can effectively denoise incoming EMG signal data [61], and the encoder-decoder model framework can be reused using a transfer learning-based model approach [56]. Once the model is trained offline using collected experimental data, the myoelectric control scheme can be readily implemented with little calibration time for the end user. Autoencoder models have already been implemented in research to improve the pattern recognition of myoelectric control schemes in the presence of electrode shift [62], but more research is needed to test the viability of using encoder-decoder networks in myoelectric control schemes.

4.2 Safety requirements in machine learning based myoelectric control systems

The active and powered upper limb exoskeletons and exosuits require high levels of safety to ensure that they do not pose any risks to human users for assistive or rehabilitative purposes. Previous research has primarily focused on incorporating safety measures in the mechanical design of exoskeletons by implementing mechanical stops, rotation limits, and force limits to prevent any excessive range of motion or force from being applied to the user [3]. However, these mechanisms may not always guarantee the user's safety when there are unknown parameter variances, hardware failures, or actuator malfunctions [63]. Therefore, control strategies that can compensate for various uncertainties and external load disturbances may significantly enhance user safety when wearing the robotic exoskeleton during tasks and movements. According to state-of-the-art research articles, one potential approach to improve safety is to apply data fusion techniques to EMG signals, considering their inherent variability arising from changes in arm posture, electrode repositing, fatigue, etc. [64]. By fusing EMG data, potential errors in motion estimation can be minimized. In [64], two data-fusing algorithms, Variance Weighted Average (VWA) and Decentralized Kalman Filter, were presented as potential methods to improve safety in robotic exoskeletons.

Additionally, other works also utilize the deep reinforcement learning-controlled neuromusculoskeletal simulator (NMMS) to validate the machine learning-based myoelectric control system. The neuromusculoskeletal simulation modulates a wide range of control schemes and parameters to test the efficacy and performance of different control methods while observing model outcomes, such as muscle force, joint kinematics and power using EMG signals [65]. Compared to the conventional control schemes of the NMMS which are broadly classified as forward-type or inverse-type. The deep reinforcement learning (DRL) based NMMS controller learns the neuromusculoskeletal system dynamics by interacting with its environment without the experimental data collected from big samples with varying anthropometrics and biomechanics characteristics. For example, [66] implemented a lower limb NMMS with DRL-based locomotion controller to validate a reinforcement learning-based myoelectric control system for a lower limb orthosis. In this work, a deep reinforcement algorithm called Soft-Actor-Critic (SAC) was used to learn the dynamics of the lower-limb NMMS and served as its locomotion controller; meanwhile, a myoelectric control scheme was trained by imitation learning through interacting with the lower-limb NMMS. Compared to validating the novel myoelectric control scheme on the human body, using the RL-based NMMS can guarantee safety while maximumly emulating the feedback from real human. However, the abovementioned research only deals with the lower-limb neuromusculoskeletal simulator, while the reinforcement learning-based neuromusculoskeletal simulator for upper limb is still extensively unexplored. To address this gap, one of our recent studies utilized the MyoSuite – a Mujoco-based neuromusculoskeletal simulation kit [67] – to simulate the flexion/extension of human's elbow joint controlled by the deep deterministic policy gradient algorithms (DDPG) – a variant of deep reinforcement learning algorithms [68]. In that work, we compared the performance between two types of action spaces – the PD-based internal model of the central neuron system, and the direct muscular activation output. The result indicated the PD-based internal model has better learning performance than the direct muscular activation output. Additionally, we also simulated the proportional myoelectric control [17] in the NMMS to validate its feasibility in validating the myoelectric control system. However, the result of muscle activation is different from the result in [17]. Therefore, further studies should focus on making the NMMS become more human-like.

4.3 Implementation of reinforcement learning algorithms

Only 3 percent of the research articles from our survey utilized the reinforcement learning algorithms for the machine learning-based myoelectric control system (**Figure 3**). However, as a branch of machine learning, reinforcement learning has some exclusive advantages if implemented in a control system. For example, reinforcement learning can inherently reflect how humans learn a skill in the real world, which is actively exploring the unknown environment and finding the long-term optimal solutions [69]. More importantly, the reinforcement learning algorithm can learn the optimal solution without the predefined knowledge about the dynamics of environment [70]. Due to these advantages, an increasing number of biomechanical studies implemented reinforcement learning, such as using reinforcement learning to control a lower-limb musculoskeletal model for obstacle avoidance [71], to control a functional electrical stimulation to assist movement [72], and to control an upper limb prothesis [73]. One literature implemented Probabilistic Inference for Learning Control Algorithm (PILCO) – a reinforcement learning algorithm to train a myoelectric control system – on an elbow exoskeleton and achieved a satisfactory result [54]. However, the PILCO algorithm depends on the dynamic model of the environment which increases the difficulty of computation and restricts the usage of trained myoelectric controllers in a single task. Different from [54], the reinforcement learning algorithm used in [72, 73] does not require the dynamic model of the environment, which called model-free reinforcement learning algorithms. The model-free reinforcement learning algorithms have the advantage of less computational difficulty, and wider applicability. There are many model-free reinforcement learning algorithms, for example, the Asynchronous Advantage Actor-Critic (A3C) algorithm [74] which was used to estimate the elbow joint torque from the surface EMG signal [75]. To explore the application of reinforcement learning algorithms in myoelectric control systems, further research is needed to validate these reinforcement learning algorithms (e.g., deep deterministic policy gradient (DDPG), proximal policy optimization (PPO), and asynchronous advantage actor critic (A3C)) used in the field of myoelectric control systems for upper limb wearable robots.

5. Conclusion

This chapter shares a review of the recent implementation of machine learning algorithms in the myoelectric control systems for upper limb exoskeletons and exosuits. The types of machine learning algorithms used in the myoelectric control systems include classification model, regression model and reinforcement learning model. Also, this chapter provides information on the methods, performance, and limitations of each myoelectric control modality. The machine learning algorithms in the myoelectric control systems have shown promising outcomes, including improved human-robot interactions, robot intelligence, and adaptiveness to the user, task and environment compared to traditional myoelectric control systems that did not use learning-based controls. Several challenges and limitations are identified which need to be addressed in future studies related to machine learning -based myoelectric control system for upper limb exoskeletons and exosuits, particularly in narrowing the gap between laboratory studies and clinical applications.

Acknowledgements

Article processing charges were provided in part by the UCF College of Graduate Studies Open Access Publishing Fund.

Author details

Jirui Fu[1], Zubadiah Al-Mashhadani[2], Keith Currier[1], Al-Muthanna Al-Ani[1] and Joon-Hyuk Park[1*]

1 Department of Aerospace and Mechanical Engineering, University of Central Florida, Orlando FL, USA

2 Department of Electrical and Computer Engineering, University of Central Florida, Orlando FL, USA

*Address all correspondence to: joonpark@ucf.edu

References

[1] Bardi E, Gandolla M, Braghin F, Resta F, Pedrocchi A, Ambrosini E. Upper limb soft robotic wearable devices: A systematic review. Journal of NeuroEngineering And Rehabilitation. 2022;**19**:87

[2] Fu J, Hosseini S, Simpson R, Brooks A, Huff R, Park J. A bilateral six degree of freedom cable-driven upper body exosuit. In: 2022 IEEE International Conference on Mechatronics and Automation (ICMA). Guilin, Guangxi, China: IEEE; 2022. pp. 617-623

[3] Gull M, Bai S, Bak T. A review on design of upper limb exoskeletons. Robotics. 2020;**9**:16

[4] Ding S, Ouyang X, Liu T, Li Z, Yang H. Gait event detection of a lower extremity exoskeleton robot by an intelligent IMU. IEEE Sensors Journal. 2018;**18**:9728-9735

[5] Wang S, Wang L, Meijneke C, Van Asseldonk E, Hoellinger T, Cheron G, et al. Design and control of the MINDWALKER exoskeleton. IEEE Transactions on Neural Systems and Rehabilitation Engineering. 2014;**23**: 277-286

[6] Massardi S, Rodriguez-Cianca D, Pinto-Fernandez D, Moreno J, Lancini M, Torricelli D. Characterization and evaluation of human–exoskeleton interaction dynamics: A review. Sensors. 2022;**22**:3993

[7] Kim J, Shim M, Ahn D, Son B, Kim S, Kim D, et al. Design of a knee exoskeleton using foot pressure and knee torque sensors. International Journal of Advanced Robotic Systems. 2015;**12**:112

[8] Fu J, Choudhury R, Hosseini S, Simpson R, Park J. Myoelectric control Systems for Upper Limb Wearable Robotic Exoskeletons and Exosuits—A Systematic Review. Sensors. 2022;**22**: 8134

[9] Dupont A, Morin E. A myoelectric control evaluation and trainer system. IEEE Transactions on Rehabilitation Engineering. 1994;**2**: 100-107

[10] Englehart K, Hudgin B, Parker P. A wavelet-based continuous classification scheme for multifunction myoelectric control. IEEE Transactions on Biomedical Engineering. 2001;**48**: 302-311

[11] Englehart K, Hudgins B. A robust, real-time control scheme for multifunction myoelectric control. IEEE Transactions on Biomedical Engineering. 2003;**50**:848-854

[12] Robertson D, Caldwell G, Hamill J, Kamen G, Whittlesey S. Research Methods in Biomechanics. Champaign, IL, USA: Human Kinetics; 2014

[13] Bi L, Feleke A, Guan C. A review on EMG-based motor intention prediction of continuous human upper limb motion for human-robot collaboration. Biomedical Signal Processing and Control. 2019;**51**:113-127

[14] Rechy-Ramirez E, Hu H. Bio-signal based control in assistive robots: A survey. Digital Communications and Networks. 2015;**1**:85-101

[15] Lotti N, Xiloyannis M, Missiroli F, Chiaradia D, Frisoli A, Sanguineti V, et al. Intention-detection strategies for upper limb exosuits: Model-based myoelectric vs dynamic-based control. In: 2020 8th IEEE RAS/EMBS International Conference for Biomedical

Robotics and Biomechatronics (BioRob). New York, NY, USA: IEEE; 2020. pp. 410-415

[16] Accogli A, Grazi L, Crea S, Panarese A, Carpaneto J, Vitiello N, et al. EMG-based detection of User's intentions for human-machine shared control of an assistive upper-limb exoskeleton. In: Wearable Robotics: Challenges and Trends. Gewerbestrasse, Switzerland: Springer, Cham; 2016. pp. 181-185

[17] Lenzi T, Rossi S, Vitiello N, Carrozza M. Intention-based EMG control for powered exoskeletons. IEEE Transactions on Bio-medical Engineering. 2012;**59**:2180-2190

[18] Roche A, Rehbaum H, Farina D, Aszmann O. Prosthetic myoelectric control strategies: A clinical perspective. Current Surgery Reports. 2014;**2**:1-11

[19] Bayat N, Rastegari E, Li Q. Human Gait Recognition Using Bag of Words Feature-Representation Method. In: Ahram T, Falcão C, editors, Human Factors and Wearable Technologies. AHFE International Conference. Vol. 29. USA: AHFE Open Access, AHFE International; 2022. DOI: 10.54941/ahfe1001481

[20] Mesin L, Merletti R, Rainoldi A. Surface EMG: The issue of electrode location. Journal of Electromyography and Kinesiology. 2009;**19**:719-726

[21] Merletti R, Farina D. Analysis of intramuscular electromyogram signals. Philosophical Transactions of the Royal Society A: Mathematical, Physical and Engineering Sciences. 2008;**367**: 357-368

[22] Li G, Li Y, Zhang Z, Geng Y, Zhou R. Selection of sampling rate for EMG pattern recognition based prosthesis control. In: 2010 Annual International Conference of the IEEE Engineering In Medicine and Biology. Buenos Aires, Argentina: IEEE; 2010. pp. 5058-5061

[23] Li G, Li Y, Yu L, Geng Y. Conditioning and sampling issues of EMG signals in motion recognition of multifunctional myoelectric prostheses. Annals of Biomedical Engineering. 2011; **39**:1779-1787

[24] De Luca C, Donald Gilmore L, Kuznetsov M, Roy S. Filtering the surface EMG signal: Movement artifact and baseline noise contamination. Journal of Biomechanics. 2010;**43**: 1573-1579

[25] Oskoei M, Hu H. Myoelectric control systems—A survey. Biomedical Signal Processing and Control. 2007;**2**:275-294

[26] Phinyomark A, Phukpattaranont P, Limsakul C. Feature reduction and selection for EMG signal classification. Expert Systems with Applications. 2012; **39**:7420-7431

[27] Lashgari E, Maoz U. Dimensionality reduction for classification of object weight from electromyography. PLoS One. 2021;**16**:e0255926

[28] Oskoei M, Hu H. Support vector machine-based classification scheme for myoelectric control applied to upper limb. IEEE Transactions on Biomedical Engineering. 2008;**55**:1956-1965

[29] Scheme E, Hudgins B, Englehart K. Confidence-based rejection for improved pattern recognition myoelectric control. IEEE Transactions on Biomedical Engineering. 2013;**60**: 1563-1570

[30] Scheme E, Englehart K, Hudgins B. Selective classification for improved

robustness of myoelectric control under nonideal conditions. IEEE Transactions on Biomedical Engineering. 2011;**58**: 1698-1705

[31] Jana M, Barua B, Hazarika S. Design and development of a finger exoskeleton for motor rehabilitation using electromyography signals. In: 2019 23rd International Conference on Mechatronics Technology (ICMT). Salerno, Italy: IEEE; 2019. pp. 1-6

[32] Arteaga M, Castiblanco J, Mondragon I, Colorado J, Alvarado-Rojas C. EMG-based adaptive trajectory generation for an exoskeleton model during hand rehabilitation exercises. In: 2020 8th IEEE RAS/EMBS International Conference for Biomedical Robotics and Biomechatronics (BioRob). New York, NY, USA: IEEE; 2020. pp. 416-421

[33] Cheon S, Kim D, Kim S, Kang B, Lee J, Gong H, et al. Single EMG sensor-driven robotic glove control for reliable augmentation of power grasping. IEEE Transactions on Medical Robotics and Bionics. 2021;**3**:179-189

[34] Su H, Li Z, Li G, Yang C. EMG-based neural network control of an upper-limb power-assist exoskeleton robot. In: Advances In Neural Networks–ISNN 2013: 10th International Symposium on Neural Networks, Dalian, China, July 4–6, 2013, Proceedings, Part II 10. Dalian, China: Springer, Berlin, Heidelberg; 2013. pp. 204-211

[35] Trigili E, Grazi L, Crea S, Accogli A, Carpaneto J, Micera S, et al. Detection of movement onset using EMG signals for upper-limb exoskeletons in reaching tasks. Journal of Neuroengineering and Rehabilitation. 2019;**16**:45

[36] Kirchner E, Tabie M, Seeland A. Multimodal movement prediction – towards an individual assistance of patients. PLoS One. 2014;**9**:e85060

[37] Wu Q, Chen B, Wu H. Neural-network-enhanced torque estimation control of a soft wearable exoskeleton for elbow assistance. Mechatronics. 2019;**63**: 102279

[38] Zeng H, Li K, Wei N, Song R, Tian X. A sEMG-controlled robotic hand exoskeleton for rehabilitation in post-stroke individuals. In: 2018 IEEE International Conference on Cyborg and Bionic Systems (CBS). Shenzhen, China: IEEE; 2018. pp. 652-655

[39] Twardowski M, Roy S, Li Z, Contessa P, De Luca G, Kline J. Motor unit drive: A neural interface for real-time upper limb prosthetic control. Journal of Neural Engineering. 2019;**16**:3-5

[40] Secciani N, Topini A, Ridolfi A, Meli E, Allotta B. A novel point-in-polygon-based sEMG classifier for hand exoskeleton systems. IEEE Transactions on Neural Systems and Rehabilitation Engineering. 2020;**28**: 3158-3166

[41] Godoy R, Dwivedi A, Liarokapis M. Electromyography based decoding of dexterous, in-hand manipulation motions with temporal multichannel vision transformers. IEEE Transactions on Neural Systems and Rehabilitation Engineering. 2022;**30**:2207-2216

[42] Tang Z, Zhang K, Sun S, Gao Z, Zhang L, Yang Z. An upper-limb power-assist exoskeleton using proportional myoelectric control. Sensors. 2014;**14**: 6677-6694

[43] Pang M, Guo S, Song Z. Study on the sEMG driven upper limb exoskeleton rehabilitation device in bilateral rehabilitation. Journal of Robotics and Mechatronics. 2012;**24**:585

[44] Loconsole C, Leonardis D, Barsotti M, Solazzi M, Frisoli A, Bergamasco M, et al. An emg-based robotic hand exoskeleton for bilateral training of grasp. In: 2013 World Haptics Conference (WHC). Daejeon, South Korea: IEEE; 2013. pp. 537-542

[45] Ngeo J, Tamei T, Shibata T, Orlando M, Behera L, Saxena A, et al. Control of an optimal finger exoskeleton based on continuous joint angle estimation from EMG signals. In: 2013 35th Annual International Conference of the IEEE Engineering in Medicine and Biology Society (EMBC). Osaka, Japan: IEEE; pp. 338-341

[46] Leonardis D, Barsotti M, Loconsole C, Solazzi M, Troncossi M, Mazzotti C, et al. An EMG-controlled robotic hand exoskeleton for bilateral rehabilitation. IEEE Transactions on Haptics. 2015;**8**:140-151

[47] Burns M, Pei D, Vinjamuri R. Myoelectric control of a soft hand exoskeleton using kinematic synergies. IEEE Transactions on Biomedical Circuits and Systems. 2019;**13**: 1351-1361

[48] Cisnal A, Pérez-Turiel J, Fraile J, Sierra D, Fuente E. RobHand: A hand exoskeleton with real-time EMG-driven embedded control. Quantifying hand gesture recognition delays for bilateral rehabilitation. IEEE Access. 2021;**9**: 137809-137823

[49] Li Z, Wang B, Sun F, Yang C, Xie Q, Zhang W. sEMG-based joint force control for an upper-limb power-assist exoskeleton robot. IEEE Journal of Biomedical and Health Informatics. 2014;**18**:1043-1050

[50] Lu L, Wu Q, Chen X, Shao Z, Chen B, Wu H. Development of a sEMG-based torque estimation control strategy for a soft elbow exoskeleton. Robotics and Autonomous Systems. 2019;**111**: 88-98

[51] Kopke J, Hargrove L, Ellis M. Applying LDA-based pattern recognition to predict isometric shoulder and elbow torque generation in individuals with chronic stroke with moderate to severe motor impairment. Journal of Neuroengineering and Rehabilitation. 2019;**16**:4-6

[52] Sierotowicz M, Lotti N, Nell L, Missiroli F, Alicea R, Zhang X, et al. EMG-driven machine learning control of a soft glove for grasping assistance and rehabilitation. IEEE Robotics and Automation Letters. 2022;7:1566-1573

[53] Meattini R, Bernardini A, Palli G, Melchiorri C. sEMG-based minimally supervised regression using soft-DTW neural networks for robot hand grasping control. IEEE Robotics and Automation Letters. 2022;7: 10144-10151

[54] Hamaya M, Matsubara T, Noda T, Teramae T, Morimoto J. Learning assistive strategies for exoskeleton robots from user-robot physical interaction. Pattern Recognition Letters. 2017;**99**: 67-76

[55] Deisenroth M, Rasmussen C. PILCO: A model-based and data-efficient approach to policy search. In: Proceedings of the 28th International Conference on Machine Learning (ICML-11). 2011. pp. 465-472

[56] Phinyomark A, Quaine F, Charbonnier S, Serviere C, Tarpin-Bernard F, Laurillau Y. EMG feature evaluation for improving myoelectric pattern recognition robustness. Expert Systems with Applications. 2013;**40**: 4832-4840

[57] Geng Y, Samuel O, Wei Y, Li G. Improving the robustness of real-time myoelectric pattern recognition against arm position changes in Transradial amputees. BioMed Research International. 2017;**2017**:5090454

[58] Tkach D, Huang H, Kuiken T. Study of stability of time-domain features for electromyographic pattern recognition. Journal of NeuroEngineering and Rehabilitation. 2010;**7**:21

[59] Boschmann A, Kaufmann P, Platzner M, Winkler M. Towards multi-movement hand prostheses: Combining adaptive classification with high precision sockets. In: Proceedings of the 2nd European Conference Technically Assisted Rehabilitation. 2009

[60] Bank D, Koenigstein N, Giryes R. Autoencoders. CoRR. 2020. DOI: abs/2003.05991

[61] Teh Y, Hargrove L. Using latent representations of muscle activation patterns to mitigate myoelectric interface noise. In: 2021 10th International IEEE/EMBS Conference On Neural Engineering (NER). pp. 1148-1151

[62] Lv B, Sheng X, Zhu X. Improving myoelectric pattern recognition robustness to electrode shift by autoencoder. In: 2018 40th Annual International Conference of the IEEE Engineering in Medicine and Biology Society (EMBC). pp. 5652-5655

[63] Kang H, Wang J. Adaptive control of 5 DOF upper-limb exoskeleton robot with improved safety. ISA Transactions. 2013;**52**:844-852

[64] López N, Sciascio F, Soria C, Valentinuzzi M. Robust EMG sensing system based on data fusion for myoelectric control of a robotic arm. BioMedical Engineering OnLine. 2009;**8**:5

[65] Zhuang Y, Yao S, Ma C, Song R. Admittance control based on EMG-driven musculoskeletal model improves the human–robot synchronization. IEEE Transactions on Industrial Informatics. 2019;**15**:1211-1218

[66] Han J, Lee J, Choi H, Kim J, Choi J. Policy Design for an Ankle-Foot Orthosis Using Simulated Physical Human–Robot Interaction via deep reinforcement learning. IEEE Transactions on Neural Systems and Rehabilitation Engineering. 2022;**30**:2186-2197

[67] Caggiano V, Wang H, Durandau G, Sartori M, Kumar V. MyoSuite–A contact-rich simulation suite for musculoskeletal motor control. ArXiv Preprint ArXiv:2205.13600. 2022

[68] Fu J, Choudhury R, Park J. Deep reinforcement learning based upper limb neuromusculoskeletal simulator for modelling human motor control. IEEE. 2023, Under Review

[69] Mnih V, Kavukcuoglu K, Silver D, Rusu A, Veness J, Bellemare M, et al. Human-level control through deep reinforcement learning. Nature. 2015;**518**:529-533

[70] Silver D, Schrittwieser J, Simonyan K, Antonoglou I, Huang A, Guez A, et al. Mastering the game of go without human knowledge. Nature. 2017;**550**:354-359

[71] Kidziński Ł, Mohanty S, Ong C, Huang Z, Zhou S, Pechenko A, et al. Learning to run challenge solutions: Adapting reinforcement learning methods for neuromusculoskeletal environments. In: The NIPS '17 Competition: Building Intelligent

Systems. Gewerbestrasse, Switzerland: Springer, Cham; 2018. pp. 121-153

[72] Wannawas N, Subramanian M, Faisal A. Neuromechanics-Based Deep Reinforcement Learning of Neurostimulation Control in FES Cycling. Italy: IEEE; 2021

[73] Pilarski P, Dawson M, Degris T, Fahimi F, Carey J, Sutton R. Online human training of a myoelectric prosthesis controller via actor-critic reinforcement learning. In: 2011 IEEE International Conference on Rehabilitation Robotics. 2011. pp. 1-7

[74] Mnih V, Badia A, Mirza M, Graves A, Lillicrap T, Harley T, et al. Asynchronous methods for deep reinforcement learning. In: International Conference on Machine Learning. New York, NY, USA: JMLR; 2016. pp. 1928-1937

[75] Wu W, Saul K, Huang H. Using reinforcement learning to estimate human joint moments from electromyography or joint kinematics: An alternative solution to musculoskeletal-based biomechanics. Journal of Biomechanical Engineering. 2021;**143**:2-3

Chapter 7

The Use of Machine Vision in the Diagnosis of Ripening Strawberries

Tamara Oleshko, Dmytro Kvashuk and Iryna Heiets

Abstract

The problem of image recognition using methods based on statistical solutions can be divided into two classes: parametric and nonparametric. In the first case there is a set of data on the selection of image criteria, in the second such criteria must be found. Taking into account, the possibilities of obtaining initial data for the identification of agricultural objects, the first option is considered in the article. Possibilities of diagnostics of strawberry growth at a stage of its cultivation in hothouse conditions are presented. Several image classifiers that can be used to solve such problems are investigated. An experiment was performed, which has shown the accuracy of the applied algorithms for diagnosis and recognition of berries, which allowed to set thresholds for the detection of ripe berries. A number of berry recognition methods were compared in order to study the accuracy of berry recognition and response to berry growth.

Keywords: image recognition, strawberries, machine vision, ripening, cultivation

1. Introduction

Image processing is an effective tool in the process of studying crops, the care of which largely depends on the observations that are made at all stages of the plant life cycle. However, when part of the crop is in the process of ripening and part is already ripe, especially in greenhouses, there is a need to control this process. Remontant strawberries, which can yield continuously throughout the year, are among the types of crops that need to be identified for ripening. Therefore, you can use machine vision to control the cultivation of such a crop. The use of such systems in agricultural work has a huge potential for accounting automation and harvesting control. Therefore, an important aspect of such tasks is the development of algorithms for recognizing the characteristics of plants and their classification, as well as the choice of existing methods.

Research in this area shows that the most common methods of machine learning for image classification include neural networks and other methods of machine learning [1–3]. However, due to the high demand for computing power, machine vision systems require both energy and time consumption for image processing. Therefore, depending on the tasks, the identification algorithm should be maximally adapted, because the criteria for selecting the quality of image recognition differ depending on the applicability to different subject areas.

Thus, to solve the narrow problems of identifying the number of ripe berries, approaches such as identifying an object by color, contour, and shape, which require only matrix calculations, can give a quite acceptable result. Taking into account that, obstacles such as brightness and image scale also play an important role, stationary camcorder installation, and artificial lighting can alleviate these problems.

The main stages, in this case, can be considered as follows:

- defining separate image contours which can be further investigated as a result of special filters application;

- determining coordinates of the selected arrays on the established color;

- determining the area of selected objects;

- setting the number of selected revolutions in order to ratio the appropriate number after a certain period of time.

Increasing the number of selected objects will be an indicator of the beginning of the harvest.

Thus, it is necessary:

- to set the number of strawberries and their area in the image for each individual area of plantings;

- to identify a separate area where video recording will take place;

- enter the number of identified ripe berries and their area together with the area identifier of individual plantations in the database in order to further verify their growth.

- video recording to implement on the basis of a moving video camera (**Figure 1**).

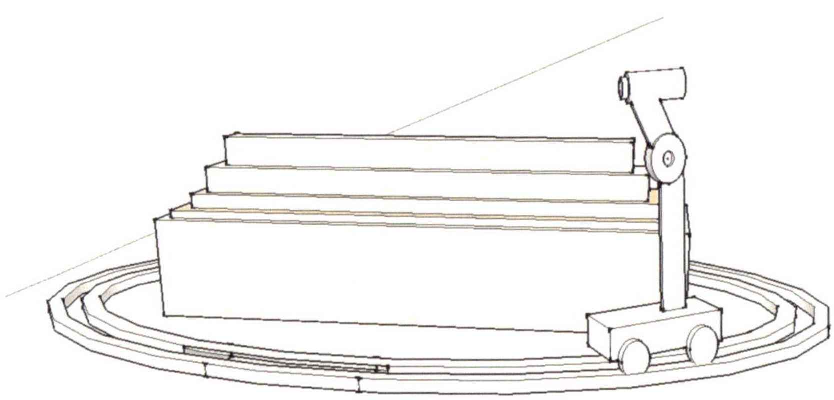

Figure 1.
Movable platform for video recording of plantings (created using Google SketchUp).

As a result, it is possible to diagnose the right time for harvesting, which will allow farmers to harvest ripe berries in time, without losses. Popular algorithms are used for image analysis: SURF; ORB; FAST; SIFT. In order to store data for further comparison, you can use a simple Sqlite database.

2. Literature analysis and problem statement

The image recognition algorithms in the field of agriculture have become very popular, which have differences and certain classifications [4], a large part of which is implemented using a number of methods of machine learning and neural networks. However, the use of neural networks requires huge arrays of data for training, which cannot always be obtained. In addition, the significant capacity for training and the time spent on processing such arrays is quite significant. It is difficult to determine the logic of the distribution of the weights rate. All this and many other requirements for big data processing impose certain restrictions on the choice of computing tools, obtaining training data, and the difficulty of understanding how the algorithm itself works. However, depending on the image identification tasks, the algorithm for determining the desired image can be simplified in some way. In particular, if you use color detectors in combination with contour detectors, for example, which allow under certain conditions to obtain image descriptors, you can identify the object with high accuracy.

There is a considerable amount of work devoted to the identification of objects in agriculture by color and geometric shapes. Thus, the recognition is performed [5] on the basis of statistical characteristics and signs of the set of obtained ranges of descriptor values with the help of the minimum distance classifier. The results of experiments on a database containing about 2635 fruits, made it possible to establish 15 different classes, which has confirmed the effectiveness of the proposed approach.

A similar study to that proposed in the article aims to analyze automated harvesting methods based on fruit categorization. It in turn offers a new approach that separates the fruit in the foreground from the background, which in addition to color spectrum characteristics also takes into account the identification of texture images. The classification model is built using reference vectors (SVM) and learns using function descriptors taken from the training data set. The proposed approach allows the use of small computing power and can run on single-board computers, such as RaspberryPi [6].

Some works describe the possibility of identifying fruits and vegetables by such criteria as the characteristics of the plant color [7], as well as the structure of their leaves [8]. Thus, the combination of these two approaches gives certain results. However, simple ways to identify images have a number of disadvantages, in particular noise generated by shadows or changes in light. Due to the use of a number of filters, the use of neural networks in this case gives a more accurate result [9]. Therefore, such filters are carefully selected. For example, the Otsu method is often used to reduce noise by low-pass filters [10].

Median filters are used to smooth inconspicuous spots [11]. Spectral analysis is used to obtain a clear identification by color. For example, researchers at the University of Oklahoma studied the use of hyper-spectral methods of image analysis to assess the quality, safety, and classification of fruits and vegetables, which resulted in determining the most effective approaches to estimating the color spectrum in the image [12]. Typically, to highlight RGB color saturation, the image classification is converted to HSV or YCbCr. To ensure image clarity, anti-aliasing filters are used, such as the Gaussian filter [13]. The step-by-step selection of individual image contours can be represented using the OpenCV image recognition library, which contains the possibilities of software implementation of the respective stages. Thus, to identify strawberries, by means of this library, you can select separate pixels by color. In other words, you can create a mask that characterizes only the berries of strawberries. Then you can reduce the noise by smoothing with the help of the Gaussian filter [13].

To determine the area of a selected object, you can count its pixel IDs, which characterize such a selection. To do this, you can use the threshold method of the OpenCV library, which turns the image, where all the pixels that are darker (less) than 127 are replaced by 0, and everything that is brighter (more) than 127 - by 255.

In this case, it is possible to set criteria for strawberry growth by comparing the specific gravity of the selected color spectrum at different times. To do this, strict conditions must be met. Photo fixation should be performed each time in the same position of the photo recorder and in the same lighting.

In this case, it is possible to observe changes in the size of the strawberry and changes in its color, which during ripening begins to correspond to the established range of identification. In **Figure 2** the corresponding range in the HSV format was set within the following limits: lower threshold—(0,50,20)–(5,255,255); upper threshold—(175,50,20) –(180,255,255).

Taking into account that the photos were taken in different conditions if error and noise in the images a3-d3 are minimal, this approach can be used to assess the growth

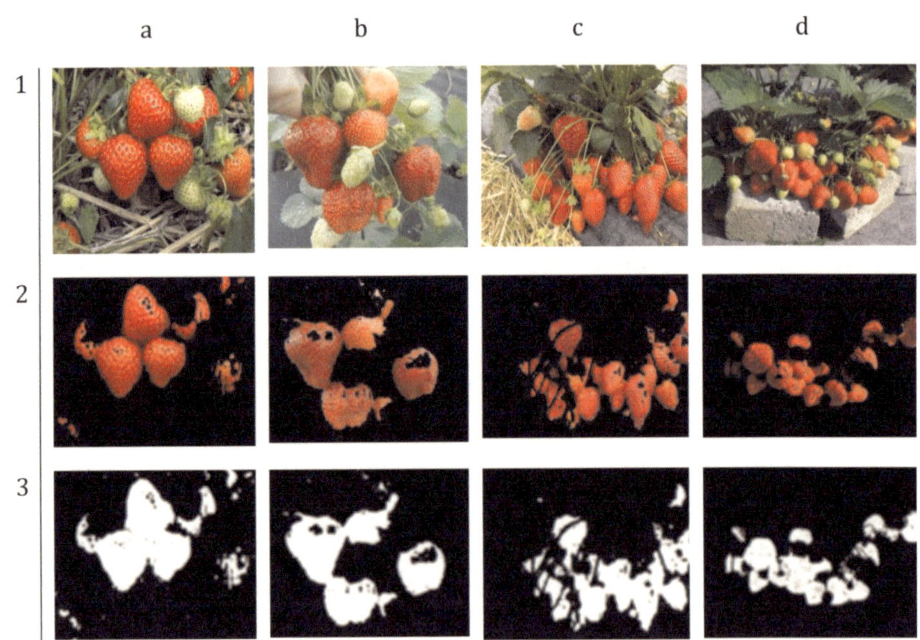

4	83923	80614	92093	63551
5	276077	279386	267907	296449

Figure 2.
The process of identifying strawberries by color.

of strawberries. In addition, if one of the most complex image processing processes is the Gaussian filter, the corresponding tasks can be solved using single-board computers such as RaspberryPi [14].

The stages of identification presented in **Figure 2** are highlighted in the following sequence:

- obtaining the original raster image from the photo-capture (.jpg or other similar format);
- color filtering;
- smoothing with a Gaussian filter and converting the selected area to black and white;
- counting the pixels that characterize the selection of strawberries (white color);
- counting other pixels (black color).

The OpenCV library was used to solve them. In the Python programming language, it looks like this:

```
import cv2
import os
im = cv2.imread('image.jpg')
im = cv2.resize(im, (400, 300))
img_hsv = cv2.cvtColor(im, cv2.COLOR_BGR2HSV)
m1 = cv2.inRange(img_hsv, (0,50,20), (5,255,255))
m2 = cv2.inRange(img_hsv, (175,50,20), (180,255,255))
mask = cv2.bitwise_or(m1, m2 )
cr = cv2.bitwise_and(im, im, mask=mask)
m1 = cv2.GaussianBlur(mask, (7,7), 3)
cr1 = cv2.GaussianBlur(cr, (3,3), 0)
print (len(cr1[cr1==0]), 'black')
print (len(cr1[cr1>0]), 'white')
cv2.imshow("msk", m1)
cv2.imshow("cr", cr1)
cv2.waitKey()
```

Thus, the ability to assess the growth of strawberries using machine vision, without using complex algorithms for deep learning and neural networks, puts forward the admissibility of our hypothesis. Is it possible to create a strawberry harvest indicator with a low-power computer and simple image operations? The main idea of this hypothesis is, taking into account the studied approaches, to obtain a growth curve of strawberries, based on a simple comparative method. Such a comparison can characterize the dynamics of changes in the area of selected objects in the images, as relative indicators of the specific weight of selected objects in the image. For example, for images a3-d3, presented in **Figure 2**, this ratio will have the following characteristics (**Figure 3**).

However, the difficulty lies in the video recording itself, as it must be carried out under certain conditions of image periodicity and with the same angle and lighting, which requires further adjustment (namely, data acquisition, storage, and subsequent processing).

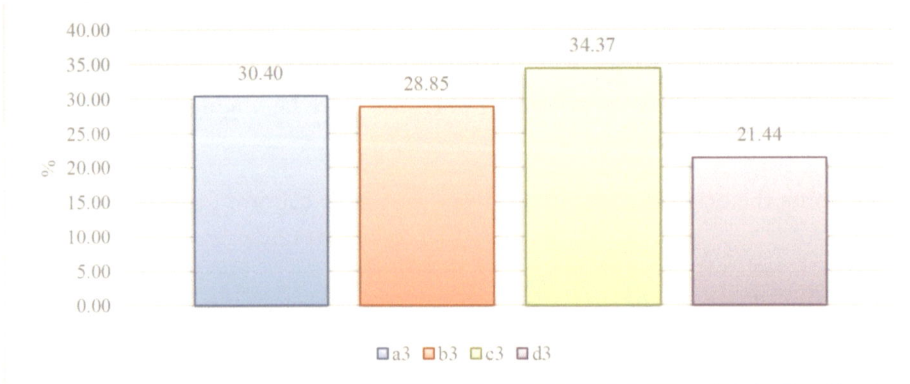

Figure 3.
Percentage of color-coded objects in images a1-d1 ***Figure*** **2.**

3. Ways to identify the growth of strawberries based on singular point detectors

In order to timely diagnose the yield of strawberries, in addition to identifiers by color, you should pay attention to other methods of pattern recognition. In the framework of the task, detecting the increase in strawberries, you can consider detectors of special points in the image, the number of which can also serve as an indicator of growth. The most common are SIFT and SURF.

The use of the SIFT detector is more efficient but requires a significant amount of computation. Its advantages are that it is possible to identify special points in the image regardless of scale and offset, as well as changes in brightness. However, the invariance of the last three transformations cannot be fully obtained. Basis for his work is the construction of the Gaussian pyramid and the differences of the Gaussians (Difference of Gaussian, DoG). Gaussian is an image that is blurred by a Gaussian filter.

$$L(x, y, \sigma) = L(x, y, \sigma) * L(x, y), \quad (1)$$

where:
L - the Gaussian value at a point with coordinates (x, y);
σ - blur radius;
G - Gaussian core;
I - the value of the original image,
* - convolution operation.

In this case, the Gaussian difference is the image obtained by subtracting each pixel of one Gaussian source image from Gaussian with a different blur radius.

$$\begin{aligned} \mathrm{D}(\mathrm{x}, \mathrm{y}, \sigma) &= (\mathrm{G}(\mathrm{x}, \mathrm{y}, \mathrm{k}\sigma) - \mathrm{G}(\mathrm{x}, \mathrm{y}, \sigma)) * I(x, y) = \\ &= \mathrm{L}(\mathrm{x}, \mathrm{y}, \mathrm{k}\sigma) - \mathrm{L}(\mathrm{x}, \mathrm{y}, \sigma), \end{aligned} \quad (2)$$

To determine a singular point, together with the construction of the Gaussians pyramid, a pyramid of Gaussians differences is constructed, consisting of the differences of adjacent images in the Gaussians pyramid. Accordingly, the number of images in this pyramid will be N + 1.

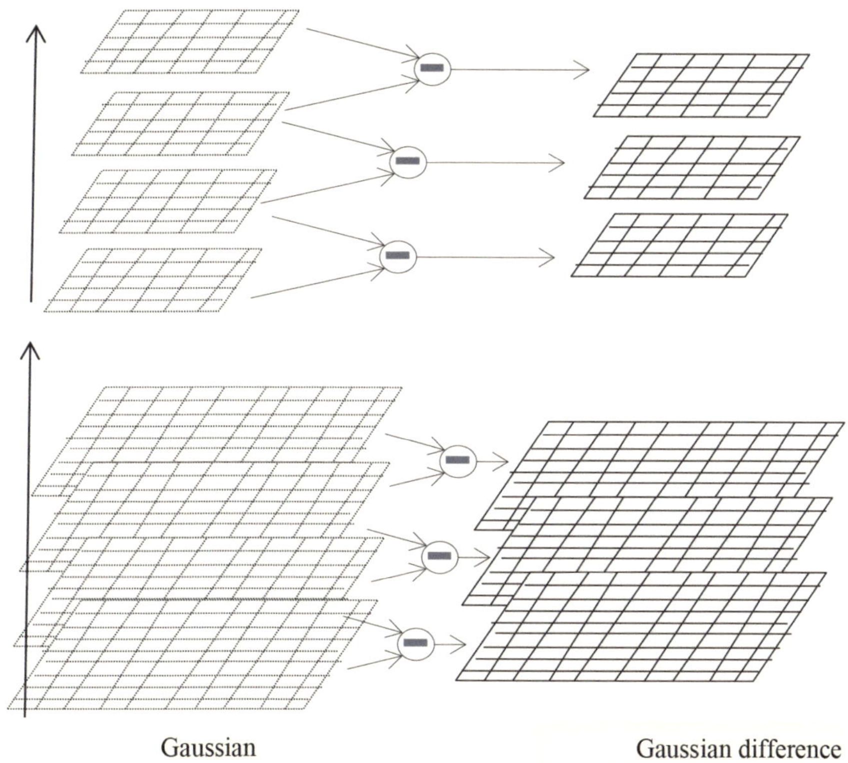

Figure 4.
Pyramid of Gaussians and their differences.

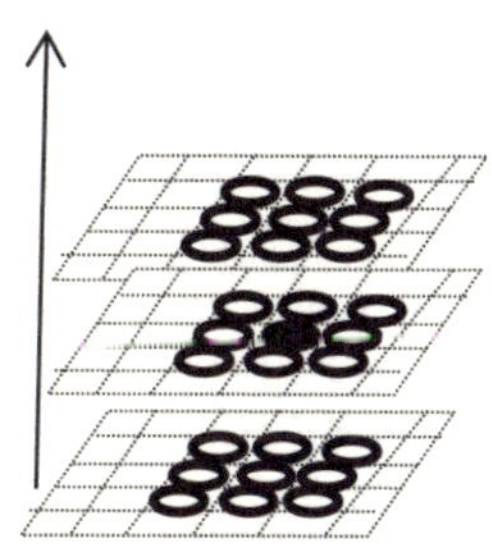

Figure 5.
Defining a singular point among the differences between Gaussians.

Figure 4 shows the Gaussians pyramid, which displays their differences, formed on the basis of two neighboring Gaussians. The number of such differences is always one less than the number of Gaussians, because when you move to the next octave, the size of the images is halved. The point that is the local extremum of the Gaussian difference is then determined. If the value of the Gaussians difference at a certain point is bigger or less than all the values of the neighboring Gaussians differences, such a point is considered an extremum point. In **Figure 5** this point is represented in shaded.

Such points are refined using an approximation of the Gaussian difference function using a second-order Taylor polynomial, which is taken at the point of a definite extremum.

Then, based on the neighborhood of the singular point, a descriptor is constructed, which is presented in the form of a gradient matrix in accordance with the pixels that are around the singular point.

Then a histogram of the descriptor is created, which is used for further correlation of the image with others. Therefore, SIFT detectors need to increase the computational speed, because the image scaling process takes some time. For greater speed, SURF detectors are used which, in order to find singular points, use the Hesse matrix, the determinant of which reaches the extremum at the points of maximum change of the brightness gradient. Thus, if the original image is given by the intensity matrix I, then the current pixel, which is analyzed for changes in color intensity, can be denoted by X = (x, y) and the scale of the filter σ. In this case, the Hesse matrix will look like this:

$$H(x,\sigma) = \begin{bmatrix} L_{xx}(X,\sigma) & L_{xy}(X,\sigma) \\ L_{xy}(X,\sigma) & L_{yy}(X,\sigma) \end{bmatrix}, \tag{3}$$

where: $L_{xx}(x,\sigma), L_{xy}(x,\sigma), L_{yy}(x,\sigma)$ - approximation convolution of the second derivative of the Gaussian cores.

The determinant of the Hesse matrix reaches its extremum at the points of maximum change of the brightness gradient. In turn, the SURF method uses a filter with a Gaussian core throughout the image, while finding singular points at which the maximum value of the determinant of the Hesse matrix is achieved. Thanks to this search, both dark spots on a white background and vice versa stand out. Unlike SIFT, the SURF method does not have a mechanism for refining points, but immediately forms descriptors. Like SIFT, SURF is invariant to rotation and scale. Using the capabilities of these detectors, you can use them to establish singular points on the structure of the image of strawberries and with the help of their number observe the yield. This statement is due to the fact that the shape of the strawberry has some common features (**Figure 6**), the identification of which with these detectors can lead to a positive result.

In order to do this, we will determine the training image, which will help to obtain descriptors of singular points (**Figure 6B**). We will also compare the number of singular points for a number of test images, which will establish which method to identify ripening strawberries is more effective, identify color changes, or identify changes in singular points. The following assumption is based on the fact that during the strawberries growth, their structure will expand, which has singular points.

Comparing these two approaches, it should be noted that the use of the SIFT detector will not allow to assess the ripening by color change, but only by size, so it has some limitations.

Figure 6.
Common features in the structure of the shape of strawberries.

According to the results of testing the training sample of descriptors of singular points, which were obtained from the image (**Figure 6b**), a small number of matches were identified on the original images using SIFT and SURF detectors (**Figure** 7). In addition, as you can see in **Figure 6**, there is a fairly large error, which indicates the inexpediency of using such detectors in the process of strawberries ripening identification.

It is presented in **Figure** 7:

1. Original test images.

2. Visual characteristics of the definition of singular points on the training image.

3. Number of singular point descriptors that coincided with the training image using the SIFT detector.

4. The total number of singular points using the SUFR detector.

5. Number of singular point descriptors that matched the training image using the SUFR detector.

6. The total number of singular points using the SUFR detector (**Figure 8**).

	a	b	c	d
1				
2	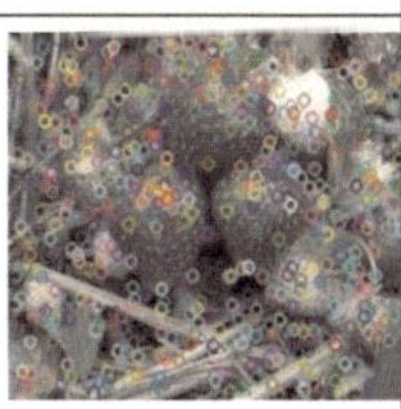	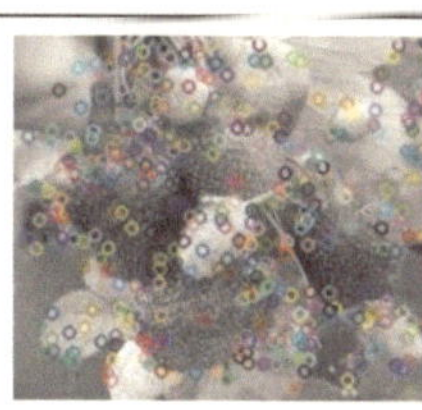	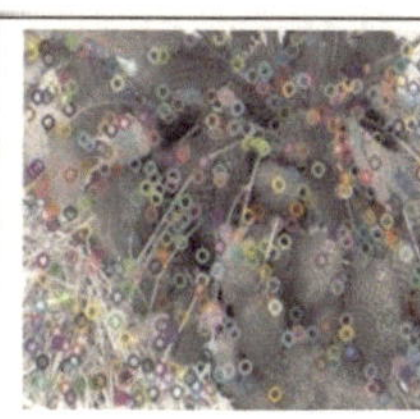	
3	37	45	47	60
4	761	520	624	1464
5	27	35	17	32
6	652	421	549	994

Figure 7.
*Visualizations and comparison of the received singular points with the training (**Figure 6b**) and test images.*

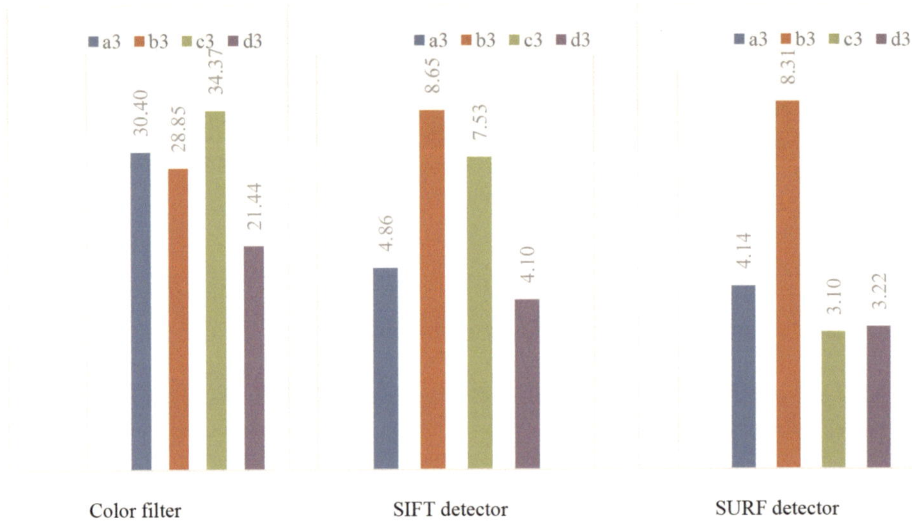

Figure 8.
*Comparison of the effectiveness of strawberry recognition methods using color filters and SIFT and SURF detectors on samples of test images (**Figure 7a1–d1**), (%).*

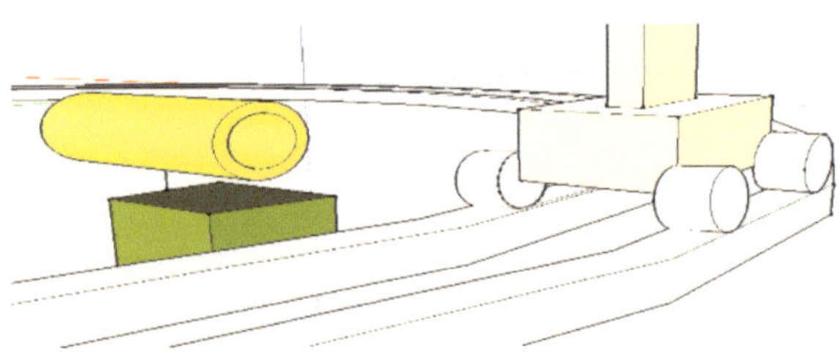

Figure 9.
Position of the infrared distance sensor (created using Google SketchUp).

4. Identification of strawberry ripening processes by color

Detection of changes in strawberry images can be implemented on the basis of a mobile platform, which is controlled by a single-board computer RaspberryPi. Infrared proximity sensors can be used to identify shooting locations, which can be used to obtain a signal to stop a moving platform at the necessary location for photo fixing of strawberries. Standard infrared distance sensors allow you to respond to approximations at a distance of up to 30 cm, which is enough to identify the stopping point of the platform. The placement of such sensors can be carried out according to the scheme in **Figure 9**.

After that, at the set time, the platform will start moving again for the next video recording.

The implementation of the software algorithm can be performed on the basis of Python interpreter and libraries: OpenCV, image processing, sqlite3, database, RPi, RaspberryPi COM port, which will be used to provide a control signal to drive the transport platform. Thus, the algorithm will be as follows (**Figure 9**).

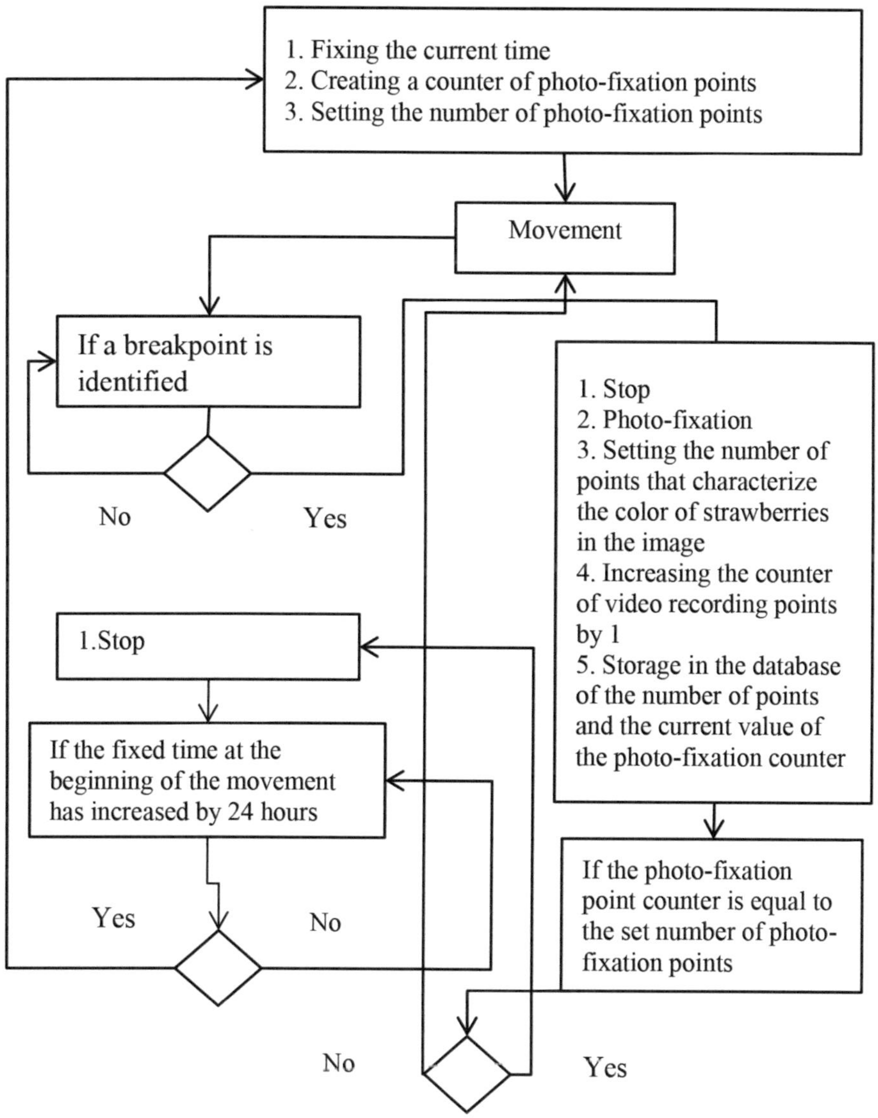

Figure 10.
Algorithm of mobile platform operation for photo-fixation of the strawberry ripening process.

By implementing this algorithm in Python [15] and using it for a test sample of strawberry images, you can get the following result (**Figure 10**). Thus, you can see how strawberries grow in **Figure 10**.

The sequence of sampling images in the order of ripening $(a_1, b_1, c_1, d_1, a_3, b_3, c_3, d_3)$ and the number of obtained points, respectively $(a_2, b_2, c_2, d_2, a_4, b_4, c_4, d_4)$. The growth dynamics of strawberries can be obtained in the form of the ratio of identified points by color to unidentified ones (**Figure 12**). However, a small percentage of errors can be observed in **Figures 11** and **12**, which are caused by the concentration of light on strawberries.

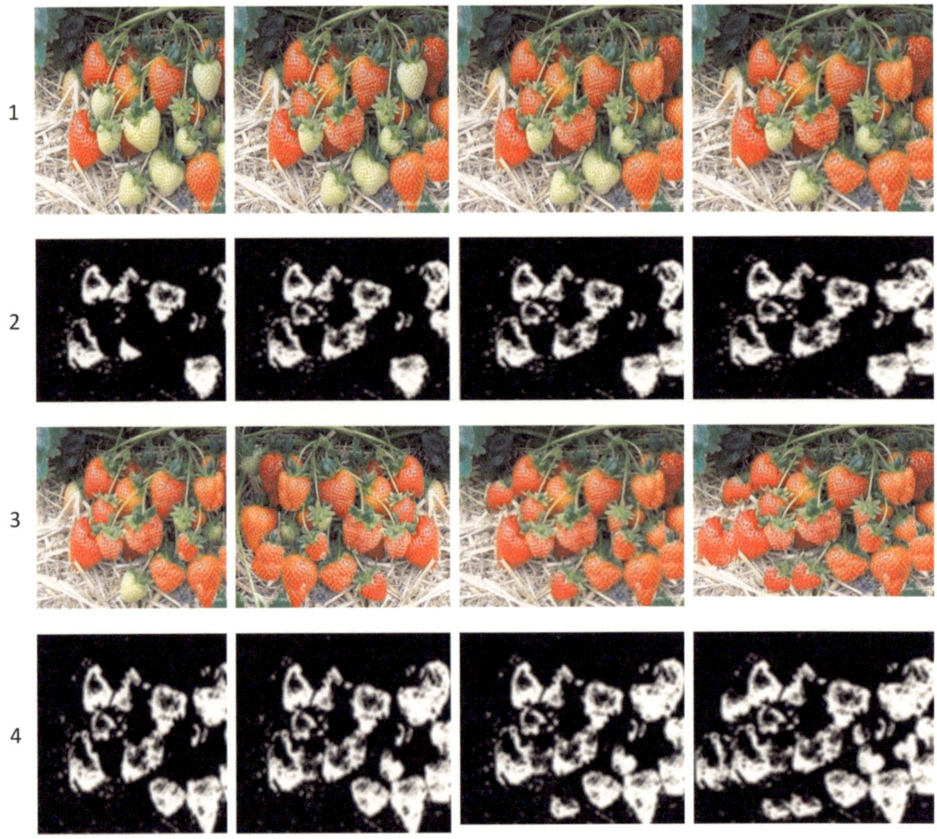

Figure 11.
Identification of strawberries by color during ripening.

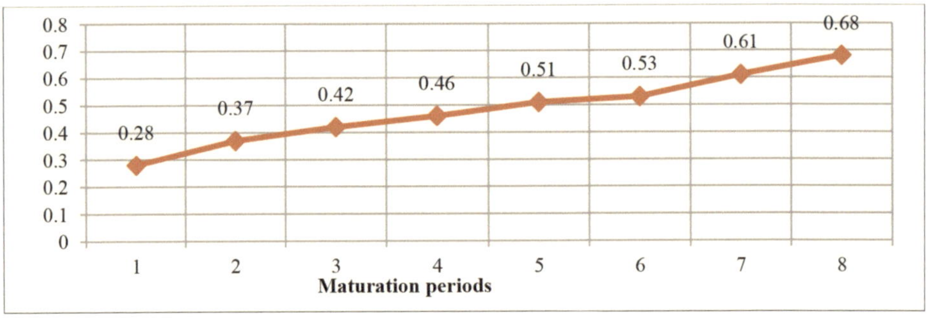

Figure 12.
The dynamics of ripening strawberries is obtained as a result of the ratio of selected points by color to the total quantity for a conditional period of time.

5. Conclusion

The most important criterion for strawberry growth is its color and size. Recognition of changes in these criteria in the images can be obtained using conventional color filters, with the specified parameters of the color range. This approach is easy to implement in practice because it does not use complex computational algorithms, which is quite important for use in agriculture.

To identify the ripening of strawberries, it is enough to use an inexpensive single-chamber computer and a mobile platform with a photo recorder, which at a given time can capture changes in the image of strawberries.

Taking into account the slight error due to the formation of the light concentration on strawberries, the experiment resulted in growth dynamics, which characterizes such an approach on the positive side. However, a number of conditions must be met for its application. In particular, detailed images require the use of means of moving the photo-fixator and the invariability of the angle of photo-fixation. This method, in contrast to the detection of singular points in the image using SIFT and SUFR detectors, has shown higher efficiency, because the most important criterion for ripening strawberries is the change of its color.

Acknowledgements

The authors thank our colleagues from School of Engineering, Aerospace Engineering and Aviation team at RMIT University (Australia) and Department of Economic Cybernetics at National Aviation University (Ukraine), who provided insight and expertise that greatly assisted the research, although they may not agree with all the interpretations and conclusions of this research.

Author details

Tamara Oleshko, Dmytro Kvashuk and Iryna Heiets*
National Aviation University, Kyiv, Ukraine

*Address all correspondence to: iryna.heiets@rmit.edu.au

References

[1] He K et al. Deep residual learning for image recognition. Computer Vision and Pattern Recognition. 2015;**2015**:770-778

[2] Sa I, Ge Z, Dayoub F, Upcroft B, Perez T, McCool C. A fruit detection system using deep neural networks. Sensors (Basel). 2016;**16**(8):1222. DOI: 10.3390/s16081222

[3] Rocha A, Hauagge DC, Wainer J, Goldenstein S. Automatic fruit and vegetable classification from images. Computers and Electronics in Agriculture. 2010;**70**:96-104. DOI: 10.1016/j.compag.2009.09.002

[4] Burgos-Artizzu X, Ribeiro A, Guijarro M, Pajares G. Real-time image processing for crop/weed discrimination in maize fields. Computers and Electronics in Agriculture. 2011;**75**:337-346

[5] Selvaraj A, Shebiah N, Nidhyananthan S, Ganesan L. Fruit recognition using color and texture features. Journal of Emerging Trends in Computing and Information Sciences. 2010;**1**:90-94

[6] Jana S, Basak S, Parekh R. 2017: Automatic fruit recognition from natural images using color and texture features. 2017. DOI: 10.1109/DEVIC.2017.807 4025

[7] Anami BS, Pujari JD, Yakkundimath R. Identification and classification of normal and affected agriculture/horticulture produce based on combined color and texture feature extraction. International Journal of Computer Applications in Engineering Sciences. 2011;**1**(3):356-360

[8] El-Helly M, Rafea A, El-Gammal S. An integrated image processing system for leaf disease detection and diagnosis. In: Proceedings of the 1st Indian International Conference on Artificial Intelligence (December 18-20, 2003). Hyderabad, India; 2003. pp. 1182-1195

[9] Kai S, Zhikun L, Hang S, Chunhong G. A research of maize disease image recognition of corn based on BP networks. In: Third International Conference on Measuring Technology and Mechatronics Automation. Shanghai, China: IEEE; 2011. pp. 246-249. DOI: 10.1109/ ICMTMA.2011.66

[10] Nunik Noviana Kurniawati, Siti Norul Huda Sheikh Abdullah, Salwani Abdullah, Saad Abdullah, "Investigation on image processing techniques for diagnosing paddy diseases", International Conference of Soft Computing and Pattern Recognition. Bangi, Malaysia: IEEE; 2009. pp. 272–277 (2009. 10.1109/SoCPaR.2009.62

[11] Ying G, Miao L, Zelin H. A study on the method of image pre-processing for recognition of crop diseases. In: International Conference on Advanced Computer Control. Singapore: IEEE; 2008. pp. 202-206

[12] Wang N, Elmasry G, Sevakarampalayam S, Qiao J. Spectral imaging techniques for food quality evaluation. Stewart Postharvest Review. 2007;**3**:1-8. DOI: 10.2212/spr.2007.1.1

[13] Haddad RA, Akansu AN. A class of Fast Gaussian binomial filters for speech and image processing. IEEE Transactions on Acoustics, Speech, and Signal Processing. 1991;**39**:723-727

[14] Gibbs S. Raspberry Pi becomes Best Selling British Computer. The Guardian. 2015. [Accessed: 28 December 2016]

[15] Kvashuk D, Oleshko T, Heiets I. Moving Platforms Source Code (Version 1.0) [Source code]. 2020. Available from: https://www.pythonanywhere.com/user/Dmitro/shares/c084641e09ba45ccbf4fb89851c121d8/